The Ten-Minute Gardener

Val Bourne has been a fanatical organic gardener since the age of five. In her twenties she had a lowly post in vegetable research, and she has always grown her own fruit and vegetables. She has a large allotment, and fruit and vegetable patches amongst her extensive flower garden in the high reaches of the Cotswolds.

Val also serves on several RHS panels, judging flower trials, and she's an Ambassador for the Hardy Plant Society. She lectures in this country and abroad, and writes regularly for the *Daily Telegraph*, *Saga*, the *English Garden*, *Garden Answers*, *Amateur Gardening* and for the Crocus website.

Her book *The Natural Gardener* won Practical Book of the Year from the Garden Media Guild in 2005 and her latest book, *The Living Jigsaw*, explains how to have a lovely garden without using chemical props, something she's pioneered.

Val has won multiple awards and was Journalist of the Year in 2014.

Also by Val Bourne

The Natural Gardener
The Winter Garden
Seeds of Wisdom
Colour in the Garden
The Living Jigsaw

The Ten-Minute Gardener's Flower-Growing Diary
The Ten-Minute Gardener's Vegetable-Growing Diary
The Ten-Minute Gardener's Fruit-Growing Diary

The Ten-Minute Gardener

VAL BOURNE

In association with
the *Daily Telegraph*

BANTAM PRESS
LONDON · NEW YORK · TORONTO · SYDNEY · AUCKLAND

TRANSWORLD PUBLISHERS
61–63 Uxbridge Road, London W5 5SA
www.penguin.co.uk

Transworld is part of the Penguin Random House group of companies
whose addresses can be found at global.penguinrandomhouse.com

Originally published in Great Britain by Bantam Press
as *The Ten-Minute Gardener's Fruit-Growing Diary* and *The Ten-Minute Gardener's Vegetable-Growing Diary*
This combined and revised edition published 2018 by Bantam Press
an imprint of Transworld Publishers

A CIP catalogue record for this book is available from the British Library.

ISBN 9781787631069

Typeset in 10.45/14.5pt Weiss BT by Jouve (UK), Milton Keynes.
Printed and bound in Great Britain by Clays Ltd, Elcograf S.p.A.

1 3 5 7 9 10 8 6 4 2

To the Best Beloved, for his helpful marginalia

CONTENTS

ACKNOWLEDGEMENTS

Thank you to Susanna Wadeson and Lizzy Goudsmit for their enthusiastic help and support, without which I would have gone under!

Thank you to Brenda Updegraff for her very necessary, eagle-eyed editing. Thank you to Andrew Davidson for the cover illustration and to Patrick Mulrey for the illustrations inside.

Thank you to Tom Poland and to Philip Lord for the design.

And thank you to my family for their patience over the long months.

PREFACE

GROWING YOUR own is a passport to paradise, because anything that goes straight from plot to plate to palate within a short time is a world away from shop-bought produce when it comes to flavour, texture and vitamin content. Health issues aside, few things in life are as pleasurable as picking sun-warmed, perfectly ripe strawberries in June, or podding peas for a summer supper. So growing your own feeds the inner hedonist in us all and that's important to our mental well-being as well as our body. On a practical note, growing your own will also save you money, and more people should invest in growing fruit, because it's one of the most expensive things on the supermarket shelves. If space is tight, you can use containers or small raised beds.

This easy-to-read handbook explains how and when to plant, what to plant and how to grow it and, whether you're an experienced gardener or a complete novice, there's plenty of expert information in this handbook. There are fruit and vegetable tasks each month, and the information on varieties is up to date. There are solutions to common problems and lots of fascinating information, so this isn't a dry book. It's meant to be a good read!

The techniques described are mostly organic because, with

fewer chemicals available to gardeners of today, we all have to be greener. In any case, a no-chemical approach seems completely sensible because we're feeding ourselves, our families and our friends and we want our food to be as healthy as it possibly can be.

Whether you're aiming to grow a few strawberries or planning to feed your entire family, this book will explain how. I write from first-hand experience, having grown my own fruit and vegetables for well over 40 years. Over the years I've made plenty of mistakes, so the advice I pass on to you is really relevant. I hope you will find this book helpful and that you gain as much pleasure from gardening and eating fresh food as I do.

Happy gardening.

Val Bourne

WINTER TASKS

FRUIT

• Plant bare-root fruit trees whenever the weather is clement. Always stake as soon as you plant and, if rabbits are a problem, use a tree guard as well.

• If bare-root plants arrive in really cold weather, find a warm spot and heel them until the weather improves.

• Prune top fruit; this includes apples, pears, quinces and medlars. Leave stone fruit (e.g. plums, peaches and nectarines) until later in the year. Apricots can be pruned in spring or late summer.

• Winter-prune autumn-fruiting raspberries, red and white currants and gooseberries.

• Check stored fruit, such as apples and pears, on a weekly basis. Remove any fruit that's gone mouldy.

VEGETABLE

To Do

Check stored vegetables
Cut pea sticks – hazel is best
Harvest winter vegetables
Complete winter digging
Spread compost
Double dig some areas
Chit early and second early potatoes
Force chicory
Clean pots and seed trays
Buy canes, string, labels, etc.
Check on your seeds
Force rhubarb
Cloche an area where the earliest sowings are to be made
Track down hibernating snails

Sow Outdoors

'Aquadulce Claudia' broad beans
'Feltham First' peas

Sow under Glass

Less hardy, taller varieties of broad beans
Lettuce

Plant

Shallot sets
Onion sets
Some garlic varieties

JANUARY

The gardening year starts here and I'm a little like Janus, the Roman god of gateways, because I look backwards and forwards at this time of year. I glory in last year's successes, that's for sure, but I try to analyse my failures too. Was it the weather, or did I choose the wrong variety? Every year is a learning curve and even experienced gardeners learn something new from year to year. It doesn't diminish my pleasure, because I always look forward to a new gardening year.

The best piece of advice for the vegetable grower is to bide your time, because being impatient and doing things too early in the year is a recipe for disaster. Timing is everything when it comes to vegetable gardening, so wait for the weather and, if you're inexperienced, ask someone you know who's clearly getting results. They'll happily share information with you.

When it comes to fruit growing, this is the month to take a long, hard look at the shape of your fruit trees, as their silhouettes are at their most prominent now that the branches are bare. Winter pruning involves shaping your trees when they're dormant and it's never drastic in most cases. Apples and pear trees are best tackled on clement January or February days. However, all highly trained fruit trees, such as cordons, fans and espaliers, should be pruned in August so that the flow of sap has slowed right down (see page 232). Any late-summer regrowth is a gentle affair, because the sap is beginning to stop flowing. Stone fruits (such as cherries, plums and apricots) are pruned lightly in summer because the rising sap seals the cuts and hopefully prevents diseases from taking hold (see page 208). It's a good idea to take a picture of these summer-pruned trees in winter, so that you can clearly see their shapes minus the foliage.

Continue planting bare-root fruit trees and bushes; if you use a fruit specialist, the range is much greater, the turnover is better and their advice and expertise is invaluable. Bare root can be planted up until late March; containerized fruit is best planted in spring.

Once the days start to lengthen, get out whenever you can with your fork and secateurs, for January is a month of preparation for the busy year ahead.

FRUIT

1 Plant Fruit Trees

(early January)

ALL FRUIT needs a warm, sunny position that is as sheltered as possible, because most fruit produces a much better crop when cross-pollinated by insects such as bees. The lure is nectar (a sugar-rich energy drink) and the flow is always better in warm conditions. This means that all fruit trees need a position that gets at least half a day's sunshine – preferably in the afternoon. Given this, the bees will visit.

Frost pockets make poor positions for fruit because blossom is inevitably browned and ruined. Observe your plot, identify the areas where frost lurks the longest and avoid using these parts. If your garden is cold (as mine is), opt for later-flowering varieties. Ideally your trees, bushes or plants should be sheltered from strong winds. Commercial orchards often use deciduous trees like willow, alder and poplar as shelter belts. Gardeners usually plant a biddable hedge (not a fast conifer hedge) or erect a fence.

The soil needs to be well drained – heavy, waterlogged soil is not good for fruit. Planting on a slope encourages better drainage, but if you are on low-lying land you may want to consider a drainage system or you could incorporate lots of grit. Most gardeners, however, can achieve good drainage by adding extra organic matter when planting. A yearly mulch also helps (see page 20).

Fruit trees are virtually permanent features of the productive garden, so it's well worth getting the planting right. You can plant in February if it is mild enough, and in March before bud break; but

October and November are the best months for planting. December is possible if the weather remains mild.

After you have chosen your site, the first job is to clear the ground of perennial weeds. It is best to hand-weed carefully while preparing the ground. Try not to use herbicide – but if you do, take care to choose one that will not damage your tree. Double digging is best as it gives roots extra depth and is especially useful where the subsoil is difficult and needs disturbing – see page 308. The addition of well-rotted compost or manure will not only improve drainage (see previous page) but will dramatically improve the soil structure.

This work can be done some time in advance, but just before planting fork in a slow-release fertilizer such as blood, fish and bone at the recommended rate.

Do not plant if the soil is very cold. You will need to keep the tree somewhere frost-free, like a shed, until it warms up. Trees come either in a pot or bare-rooted. Generally, it is better to buy bare-rooted specimens because they adapt better to your soil conditions – for concise tips on how to plant them, see page 285.

Dig a hole wide enough to take the roots of your tree fully spread out. Mound the soil up slightly in the middle. Make sure that you plant the tree to the same depth as before, which should be visible as a mark on the stem. If you can't see it, make sure that the graft scar, if there is one, is about 10cm (4in) above the finished soil surface. Back-fill the hole, working the soil into any gaps, and finally firm and level the soil. Stake straight after planting, making sure that the trunk is about 8cm (3in) from the stake. Add a mulch of 5–8cm (2–3in) of compost or manure if the ground is warm enough (see page 20). Fasten the stem to the stake using a soft tie, then water well.

If you're serious about soft fruit you will need a good fruit cage to keep the birds away. Redcurrants, raspberries and blueberries are top targets – and who can blame them? Fruit cages are not cheap, but you will recoup the cost relatively quickly because fruit is so expensive to buy. If a metal one is out of the question due to price, improvise and build a wooden one.

Did you know? Fruit varieties are often over 100 years old and many are local to a particular area. In many cases they have adapted to local conditions like poor soil, a windy location or cold weather. Research your local varieties by consulting a specialist nursery, because regionality is really important when it comes to apples and pears. Local apple groups and Common Ground (www.commonground.org.uk/ 01747 850820) will advise you.

SECRETS OF SUCCESS

- Research thoroughly. Look round the gardens in your local area to see what types of fruit grow well.
- Explore varieties and choose those that suit your site. Select the correct rootstock (see page 207) to go with the variety.
- Take into account flowering times and self-incompatibility when selecting varieties (see page 107).
- Prepare the soil well when planting all fruit.
- Feed regularly as appropriate.

Organic Tip ✓

Don't just buy any old apple tree from the garden centre. Order fruit from a reputable fruit specialist so that you get healthy certified stock. Seek their advice. Avoid family trees bearing three or more varieties. They sound fun, but one variety always wins out at the expense of the others, leading to an ugly tree.

2 Winter-prune Gooseberries

(early January)

GOOSEBERRIES may be savage, prickly beasts, but they provide the first fresh fruit of the year for many gardeners. Culinary varieties can be cooked alone, or added to mixed fruit compôtes. Some dessert varieties can be eaten raw, but unfortunately their distinctive muscat flavour is adored by blackbirds. You can buy standards (bushes grafted on to a tall single stem) or bushes; both produce a large amount of fruit – up to 5kg (11lb) per bush in some cases.

Gooseberries are very versatile. They are self-fertile, so one bush can be grown alone, which is an advantage in a small garden. Gooseberry bushes can also do well in partial shade and they withstand harsh conditions, of both temperature and wind, which makes them a good choice for cooler areas – though if you have a windy garden you should probably avoid standards, as they can snap under the weight of fruit in summer gales. Standards do fit well into flower borders, however, and this is one fruit bush that is extremely long-lived.

In winter cut back the leading stems by half. The side shoots (or laterals) are cut back to two buds. This encourages fruiting spurs to form on the old wood. Once this has been done, check that the framework of the bush is open enough and remove more wood if the crown seems congested. An open middle makes picking easier.

Did you know? Lancashire weavers were a competitive bunch and by the 1740s they were holding gooseberry shows where the biggest and best won. By 1815 there were 120 of these shows in the north of England. The largest yellow gooseberry, exhibited in Stockport in 1830, was named 'The Teaser'. The largest red, from Nantwich, was dubbed 'Roaring Lion' and the largest white, from Ormskirk, was 'Ostrich'. Darwin wrote about their 2oz fruits and many varieties still grown today date from this competitive era.

Organic Tip ✔

Mildew is a disease of drought-stressed plants and can be a problem with gooseberries, especially on light soil or in drier areas. American gooseberry mildew arrived in the UK in the early 1900s and devastated the gooseberry industry. However, if mildew does strike, it isn't fatal and modern varieties have been bred for resistance. If you are on the eastern, drier side of Britain, or if your soil is light, always opt for a modern variety and mulch in summer.

SECRETS OF SUCCESS WITH GOOSEBERRIES

- These versatile plants are easy to grow and they thrive all over Britain in a variety of soils and situations, including cold gardens and semi-shade.
- The culinary varieties are too sour to tempt the birds, but the dessert varieties come under concentrated attack in July. These will need netting.
- Dig in 5cm (2in) well-rotted manure or garden compost over the whole area before planting.
- Plant each bush 1.5m (5ft) apart during the dormant season (November–February).
- Stake half-standards well and tie them in using soft ties.
- If you are trying to produce large dessert gooseberries to eat, thin the fruit in May by removing every other berry. The thinnings can be cooked.

VARIETIES

'Invicta' AGM
This recent, vigorous culinary variety is immune to American gooseberry mildew. It yields very highly. Pick in late July and allow 1.8m (6ft) between each bush.

'Hinnomaki Yellow'
This dessert variety produces large, yellow fruit similar to the popular old but mildew-prone variety 'Leveller'. Pick in mid-July. It is one of a Finnish-bred series, so is very hardy.

'Pax'
A virtually spineless, upright gooseberry with large, plum-red berries that can be cooked or eaten. Mildew resistance is very high. Pick in July.

'Whinham's Industry'
This dark-red, hairy gooseberry has a superb, very sweet flavour and makes wonderful jam. Raised by Mr Whinham of Morpeth before 1850, it produces a remarkable crop.

3 Winter-prune Apples
(mid-January)

NOW IS the time to winter-prune your dormant apple trees using sharp secateurs and a pruning saw. Winter-pruning should be a gentle affair carried out in clement conditions, not in really cold or unpleasant weather. Remove the 3 Ds – dead, diseased or dying wood. Then take out any branches that cross and any that are too close to the ground. The aim is to produce an open-centred tree, but remember that the more you prune, the more the tree will try to replace the leafy growth it has lost and the less fruit it will produce. Varieties vary greatly in habit and bud formation, but try to encourage branches to develop horizontally and cut out those that rise up steeply. More fruit buds are formed when the sap is running more slowly. All cuts should be made just above an outward-facing bud, if possible, to encourage a spreading habit.

How you prune will depend on the maturity of the tree, but all winter pruning should aim to preserve the fruit buds. These are fatter and rounder than the pointed leafy buds and it should be possible to spot them on short, knobbly spurs and on the year-old wood of lateral branches. The object is to preserve these and encourage more spurs and laterals to develop. Be aware, though, that a few apple varieties are tip-bearing (with apples forming on the ends of shoots). With these, all the unprofitable wood should be removed and the laterals should be shortened to four–six buds.

Did you know? Until recently it was thought that apples had evolved from the wild crab apples found in the English countryside. However, new genetic research has revealed that our apple varieties began life in the fruit forests of the Tien Shan in central Asia. Initially seeds were spread by bears and birds, but travellers and their pack animals also ate the fruit and so spread the seeds beyond their native region.

VARIETIES

'Lord Lambourne' AGM
A compact dessert apple that's excellent in a small garden. Aromatic flavour and good Cox colour. Pick in mid-autumn and store until late autumn. Pollination Group B.

'Egremont Russet' AGM
Honey-coloured, rough skin and grainy flesh with a distinctive nutty flavour. Pick in mid-autumn and store until late autumn. Pollination Group B.

For further varieties of apple, see September, pages 260 and 263.

Organic Tip ✔

Always use sharp secateurs and pruning saws when pruning fruit. A sharp, clean cut allows the wound to callous over naturally. Damaged, frayed cuts allow disease to enter.

SECRETS OF SUCCESS

- Aim to end up with an airy shape that simplifies the silhouette. Try to achieve a goblet shape with about five main branches.
- Teach yourself to recognize fruit buds as opposed to leaf buds.
- Gentle pruning encourages gentle growth and that will encourage fruit buds.
- Hard pruning provokes a quick reaction. In fruit this means lots of slender, fast-growing shoots that bear leaves but not fruit.
- Growth is controlled by rootstock. Dwarf rootstocks need light pruning. See page 234.
- Some varieties, such as 'Rev. Wilks' (an excellent early cooker), have exceptionally spreading habits, so the pruning of young trees should be to inside and not outside buds as is usual.

4 Winter-prune Pears

(mid-January)

PEARS PREFER warmth, so are generally less suited to the cool British climate than apples. Certain warmer areas of Britain, often close to rivers, do suit them, however, and Waterperry in Oxfordshire has the name to prove it. It isn't just a north–south divide, though. The area round Jedburgh in south-east Scotland was at one time a renowned pear-growing district. However, growing pears in some British gardens can be difficult and this is why pear trees were often planted on warm walls in centuries past. Warmth is also needed for the growth of the pollen tube before fruit can be formed (see page 107), so a cool early summer can affect the crop. Pears usually give lower yields than apples and they do not store as reliably either.

Many pears form upright, goblet-shaped trees with a narrow arrangement of tall branches. Winter pruning will keep the shape open and the leaders can be tipped back every winter. These upright branches can be cut diagonally down from left to right in one year and then from right to left in the following year to balance the shoots and keep the leader straighter. The laterals and side shoots are shortened, as they are for apples, to four–six buds.

Quince is used as a rootstock for pears because they are closely related, and as a result they are thirsty. There are two types: Quince A rootstock, which is vigorous, produces trees that reach up to 3–9m (12–19ft) or more, while the less vigorous Quince C trees reach up to 3m (10ft) after 5 years. Fruit-growers graft vigorous varieties on to Quince C to slow their development, while less vigorous varieties are grafted on to Quince A to boost their growth.

Quince C is suited to wetter soil, but Quince A is the most widely available. (For more on rootstocks, see page 207.)

Did you know? By the twelfth century pears were being imported from France and many of today's pear varieties have French names like 'Glou Morceau'. Over time, texture changed from gritty pears that often needed cooking (like 'Doyenne du Comice') to butter-textured pears that were delicious eaten fresh. However, in 1885 a new pear called 'Conference' was bred by the Rivers nursery in Sawbridgeworth and it soon became the most successful commercial variety in Britain.

Organic Tip ✔

Pears are naturally very deep-rooted and this balances their tall physique. However, because they are grafted on to quince rootstocks, which are shallow-rooted, this makes them vulnerable to being rocked by the wind when in leaf. Therefore pear trees need staking for at least 5 years – much longer than apples.

SECRETS OF SUCCESS

- Pear blossom opens earlier than apple blossom and is more likely to get frosted, so choose a warm position in the most frost-free place you have.
- If your garden is cold or frost-prone, opt for a later-flowering variety. Cold temperatures and pear production do not go together.
- Choose a good variety from a specialist.
- Stake for 5 years.
- Take time and trouble when pruning in winter.

VARIETIES

'Doyenne du Comice' AGM
The classic, grainy-textured pear for cooking – a plump, round, pale-skinned delight. Needs a warm position. Pick mid-autumn and store until early winter. Scab can be a problem. Pollination Group C.

'Conference' AGM
The most reliable pear of all, producing slender fruit for eating rather than cooking. Pick when firm in early autumn and allow to ripen. Keeps for 4 weeks. Self-fertile, but fruits best when cross-pollinated. Pollination Group B.

'Concorde' AGM
A British-bred, self-fertile hybrid between 'Conference' and 'Doyenne du Comice', bearing heavy crops of medium-sized, rounded fruits on a compact tree. Pick in late October. Pollination Group C.

'Beth' AGM
A new English-bred variety that comes into fruit early and regularly. The white flesh has a melting texture and yields are high. Pick in late August. Pollination Group D.

For further varieties of pear, see September, page 260.

5 Renovate Old Fruit Trees

(late January)

IF YOU HAVE inherited an old fruit tree and it's cropping well, leave it alone. However, if it's in a bad condition but you enjoy the fruit, it may be worth taking the time and trouble to rejuvenate it. To lessen the shock, this should be tackled over 4–5 years rather than all in one go. If the experiment fails, console yourself with the fact that fruit wood makes wonderful logs.

The project is often more viable with an apple or pear tree than with a stone fruit tree, but if you love your tree, do fight for it. Tackle one quarter per year during clement winter weather. Aim to simplify the shape so that the canopy is open, allowing air to circulate. A shaded middle is bad news. Shade prevents the initiation of fruit buds because the wood doesn't get chance to ripen in the sun: it tends to become limp and etiolated instead. Any fruit produced is likely to lack flavour too because the sun can't ripen it. Good light is also vital for sweet fruit.

Start by removing all dead, diseased and dying wood, then cut away any crossing branches right across the tree. This will simplify the shape and allow you a better view. Examine a quarter of the tree from afar and decide where to make the cuts that will improve the shape. Large limbs may need removing and you may have to cut them away in sections. You will need to make undercuts (from the bottom of the branch) first before sawing downwards so that you get a clean cut. Aim to saw back to where the branch you are removing joins a larger branch. Try to cut it almost flush, but leave a stub to callous over. Do not leave a stump.

SECRETS OF SUCCESS

- Treat every tree as an individual.
- Be safe. Invest in a good stepladder or platform.
- Start by winter-pruning and tidying the entire tree, then assess the shape of one quarter before radically pruning it, one quarter per year.
- Use sharp tools and make undercuts before sawing downwards.
- Take your time and get down to the ground regularly between cuts to assess the shape of your tree.
- Don't leave stumps – they will die back and cause fungal diseases.
- Tidy up thoroughly afterwards.

Did you know? The original 'Bramley Seedling' still survives in Church Street, Southwell, Nottingham more than 200 years after Mary Ann Brailsford planted a pip in her parents' garden at some point between 1809 and 1813 when she left to get married. In 1846 the house was bought by a butcher named Matthew Bramley. In 1856 the tree came to the attention of a local nurseryman, Henry Merryweather, who asked if he could take cuttings. Matthew Bramley agreed but insisted that the cooking apple should be named after him. The first trees descended from it were sold in 1862. In 1900 a severe storm blew down the original Bramley apple tree, but it was restored to health and continued to bear apples.

In 2016, scientists recorded that it is dying.

Organic Tip ✓

Bear in mind that the majority of fruit trees are most productive up to 25 years of age, so a tree older than this will never be as bountiful as a youngster. On the plus side, a mature, gnarled fruit tree is an amazing sculptural feature and supports a lot of wildlife.

6 Mulch and Feed Fruit Trees

(late January)

MULCHING any plant has to be carefully timed because the ground must be moist and reasonably warm; otherwise you could trap frost into the soil for some months. Towards the end of January winter often begins to loosen its grip and mulching now will retain moisture in spring and suppress weeds. Both will help your fruit trees, bushes and canes greatly. Care must be taken to keep the mulch away from the trunks of trees, however.

Over time, a mulch of organic matter will be pulled down into the ground by worms and this will improve the soil structure and aid drainage. The very best material for mulching fruit trees is well-rotted animal manure – this should smell sweet, without any hint of ammonia. Source it carefully: some gardeners have had problems with manure contaminated by a persistent hormone weedkiller called aminopyralid (see page 284). For this reason, many people have turned to using their own garden compost instead.

Some gardeners prefer decorative mulches like bark, especially if their fruit trees are in the garden. All mulches rot down on the soil's surface, using up nitrogen in the process. If the material is low in nutrients (like bark, for instance), you need to replace lost nitrogen by sprinkling on a nitrogen-rich feed before mulching. Powdered chicken manure (sold as 6X) is light to handle and an excellent plant food for leafy growth. A further application of fertilizer in March is also useful.

Did you know? Garden soils generally contain high levels of phosphate, so bonemeal should be used only every 3 years. Apply 100–125g per square metre (3–4oz per square yard).

Organic Tip ✔

Blackcurrants, pears and plums need more nitrogen than other fruits and these should be given a nitrogen feed every March: 100g per square metre (3oz per square yard) is ideal.

SECRETS OF SUCCESS

- Try to use organic fertilizers. They improve soil structure as well as adding nutrients. They include manure, slurry, seaweed, guano, compost and bonemeal.
- Liquid seaweed seems to cure most deficiencies and it helps fruit to resist pests like red spider mite, aphids and fungal infections.
- Always follow the instructions on fertilizer packets to the letter. Weigh it to ensure you use the right quantity.

SOIL REQUIREMENTS

Nitrogen
This promotes leafy growth rather than flower. As a result, crops like strawberries, raspberries and apples do not benefit from too much nitrogen.

Potash
This is essential for fruit colour and flavour, for flower-bud development and for hardening the wood to prevent frost damage. Red-edged leaves indicate a shortage. It should be applied once a year in spring. Use sulphate of potash or seaweed meal as directed. Crops like strawberries need a high-potash feed every 2 weeks during the growing season. Tomato food has just the right proportion of nutrients for fruiting and flowering plants. This is watered on.

Magnesium
This is usually deficient only in sandy soils. The biggest clue is early leaf fall. Epsom salts can be applied: 33g per square metre (1oz per square yard).

Iron
Alkaline soil is often short of available iron and raspberries are prone to iron-deficiency, which causes them to develop yellowing foliage. Use a water-on iron supplement, or mulch heavily to reduce soil pH, so that any iron in the soil then becomes available.

VEGETABLE

1 Buy and Chit Early Potatoes

(early January)

THIS is the time to buy your seed potatoes because the really popular varieties tend to sell out. All potatoes are labelled as either first early, second early or maincrop (according to when they crop) and the tubers are guaranteed to be disease-free by the growers. The first and second early varieties should be allowed to 'chit', or produce shoots, before planting. Remove them from the packaging – taking care to wash your hands afterwards because many tubers are treated with fungicide – and lay them out on large egg trays or on clean seed trays. Look for the small eyes (the indentations with tiny buds) and angle each potato so that one eye is facing upwards. Slowly (as the light levels rise) the buds will develop into small, sprouting shoots. These race away once the tubers are planted, helping the crop to develop more quickly. It isn't necessary to chit maincrop varieties, but chitted early potatoes do produce a heavier crop.

Store the trays of potatoes somewhere cool. A garden shed is ideal. However, the tubers must be kept frost-free and out of the reach of mice and rats. So make sure that your shed is secure and well insulated.

Chitting must be done in a cool position so that the sprouting growth stays compact. Too much warmth produces long, etiolated shoots that snap off as you plant. Plant them outside from mid-April onwards.

Did you know? The potato is high in vitamin C and it was a principal source during two world wars, so much so that when the crop failed in 1916 scurvy broke out among the army. A medium helping of new potatoes will give you half your daily vitamin C allowance.

Organic Tip ✔

Do grow your own. Commercial crops of potatoes are frequently sprayed against blight. The 'Sarpo' varieties (bred from wild solanum species originally collected by the Russian geneticist Nikolai Vavilov around 1925) are selected for total blight resistance. First and Second Earlies usually miss blight.

SECRETS OF SUCCESS

- Gamble and plant four or five tubers now, protecting them with plastic or glass cloches. Plant a few earlies in March (see page 98), keeping the rest of your chitted tubers until mid–late April.
- Potatoes are very susceptible to frost damage. Earth your early potatoes up (i.e., mound soil over most of the emerging growth), covering the new foliage, to protect the vulnerable shoots. If a cold night is forecast, fleece your crop.

FIRST EARLY VARIETIES

'Foremost' AGM
White, firm, oval tubers.

'Red Duke of York' AGM
A red-skinned, floury, yellow-fleshed potato. Best steamed.

'Accent' AGM
Yellow, round potatoes. Excellent eaten cold.

SECOND EARLY VARIETIES

'Lady Cristl'
Bulks up very well. Always early. Oval, creamy tubers with a firm texture. Eel-worm resistant.

'Nadine'
A handsome, waxy, creamy potato. Good to eat and exhibit.

'Charlotte'
Yellow-skinned, waxy tubers. First-rate flavour.

'Belle de Fontenay'
Also planted as maincrop. A heritage French creamy potato producing dogleg-shaped tubers. Bakes brilliantly, but susceptible to blight.

For further varieties, see August, page 242 and September, page 281.

2 Double Digging

(early January)

CROPS DO best on well-fed garden soil. The compost from your heap is airy and fertile, and this is the time to dig it out. However, most heaps never get hot enough to kill off all the seeds that inevitably congregate. When the compost is spread over the soil surface, seeds germinate and, sadly, they are normally weeds. So the best way to incorporate well-rotted compost and avoid the problem of weeds appearing is to bury it a spit (or a spade's depth) below the soil surface. This is called 'double digging' and the words do sound ominous.

However, you will only need to double dig each area once every three to four years, and in any case by January most gardeners are normally straining at the leash to get outside and take some exercise.

You will need a sharp spade, a fork, a wheelbarrow, a line and a groundsheet. A stout plank to stand on, so that you don't compress the soil, is also an essential.

Section off a 2.4m by 1.2m (8ft by 4ft) area of garden. Lay the sheet down on the left-hand side if you're right-handed. With your spade, remove a spit of soil from the entire area and heap it up tidily on the sheet. You will end up with a neat, flat-bottomed trough in the ground and a pile of soil by the side. Take your fork and break up the ground at the base of the trench, then incorporate your garden compost, well-rotted manure or a mixture of both into the ground. Then replace all the soil to form a flat-topped mound. The decomposition process carries on, warming the soil and helping your crops to grow. The '8 by 4' bed is small enough to make double digging, and planting, quick and easy.

Did you know? For centuries horse dung was always considered to be the finest animal manure. Arab gardeners, who first used hotbeds over 1,000 years ago, fed their best horses with a special diet of barley, beans and alfalfa to produce superior, nitrogen-rich manure.

EQUIPMENT

A stainless-steel spade, a strong fork and a line for measuring.

A decent, light wheelbarrow.

A waterproof groundsheet and wellingtons.

Regular cups of tea or coffee, hopefully provided by someone who will admire your work.

Organic Tip ✓

Certain substances speed up the composting process by providing a nitrogen boost. A layer of nitrogen-rich chicken, hamster, rabbit or guinea pig droppings works well. Comfrey leaves also act as an accelerator; the clone 'Bocking 14' is the best at offering balanced nutrients. Chop up the leaves and make a leafy layer on top of your compost every month. Another system is to use urine, but the tried-and-tested method of peeing on your compost heap is far more attractive to the male gardener, and logistically easier too!

SECRETS OF SUCCESS

- You can double dig only in good weather when the ground is on the dry side. Then every spade of soil is lighter.
- A good compost heap with a lift-off front is essential so that you can remove a wheelbarrow of compost easily. Don't overfill the barrow.
- Pace yourself – there is no rush – and think of the increase in yield that your double digging will bring about.

3 Winter Care of Brassicas

(mid-January)

BRASSICAS are the gardener's standby in winter. They provide months of food, but they do need the cold weather to develop their distinctive sweet flavour. You can emulate the cold process by picking Brussels sprouts (or any other brassica) and storing them in the fridge for 10 days.

The hardiest of all are probably the kales, which will survive even in savage winters. The green curly kale can be picked from October through to spring and there is also a red version, 'Redbor'. The slender-leaved Tuscan varieties, like 'Cavolo de Nero', are just as hardy as the others. Red cabbage will also shrug off the weather well. These toughies should always be grown.

Brussels sprouts and purple sprouting broccoli can succumb in hard winters, so tidy up all fallen leaves now because they can harbour disease and shelter slugs. Remove any cabbage stumps, or any Brussels plants that have finished cropping. If you haven't netted against pigeons, do so now. They will be at their most voracious in March, just when the purple sprouting broccoli is budding up. You may have to knock snow off the net.

All brassicas require a great deal of nitrogen. It's traditional to plant them in a plot that was occupied by legumes the year before, because the nodules on the roots of peas and beans fix nitrogen into the soil naturally. Onions and shallots usually follow brassicas and these shallow-rooted, bulbous vegetables are also quite greedy feeders, so dose the soil with blood, fish and bone (a slow-release fertilizer; see page 317) as you remove old brassica plants.

Did you know? The Brussels sprout was first recorded in Brussels in 1750 and is thought to have been a natural hybrid related to wild kale. Sprouts became popular in Britain about 100 years later, but the tall varieties bred in the late nineteenth century have been superseded by shorter modern F1 hybrids that button up all along the stem.

SECRETS OF SUCCESS

- See page 124.

Organic Tip ✔

Perhaps the most widespread predator of the Cabbage White caterpillar is a tiny wasp called Cotesia glomerata. *It lays eggs into the caterpillars as they emerge from the egg cases. The wasp larvae develop inside the growing caterpillars and eventually burst out of their bodies and almost immediately form sulphur-yellow cocoons. These resemble yellow rock wool and this can be seen in nooks and crannies – often in the greenhouse. Should you spot any, leave the cocoons to develop.*

VARIETIES

Kale 'Dwarf Green Curled'
Very crinkled leaves on compact plants (up to 40cm/16in). A superfood, although once despised as a poor man's crop. October–April.

Red Cabbage 'Red Jewel' F1 AGM
Large, tightly packed hearts of crisp, ruby-red leaves. Stands and stores well. Cut after Christmas.

Kale 'Cavolo de Nero'
Slender, dark-green leaves that can be harvested from late autumn right through the year. Handsome on the plot.

Brussels Sprout 'Bosworth' AGM
Mid-season sprout with firm, smooth, dark-green buttons well spaced on the stem. Stands well in winter.

4 Prepare for Seed-Sowing

(mid-January)

WASHING pots and seed trays could never be described as exciting. It's definitely a job for a bright day and the radio does help. But hygiene is important in seed-sowing and this is one of few times of year when you can tackle it.

Brush the debris off trays and pots. Wipe them over with a damp cloth and immerse them in a bowl of hot water, then take up the scourer. Allow yourself a dab of washing-up liquid. Lay the trays outside on a bright day, rinse them with a hose and let them drain and dry before stacking them.

If investing in some new pots or trays, choose smooth-sided and -bottomed ones with no indentations. Make sure there is no lip on the edges. Trays like this provide little or no opportunity for slugs to sleep under or near your seedlings.

Once you start sowing, always water seedlings with mains water using a can with a fine rose. Have at least two cans if possible and when you empty one, fill it up and allow it to stand for several hours. This will warm up the water and allow chlorine to escape. The rose should be facing upwards to allow the water to fall as a fine rain. Try to water before midday, ventilate the greenhouse or shed, then fill both cans and shut them inside.

Did you know? 'Watering pots', usually with handles on the top, were used by gardeners for centuries. The earliest known appearance of the term 'watering can' was in the 1692 diary of the keen Cornish gardener Lord Timothy George. In 1886 the Haws company patented a new design with a single round handle at the rear. This soon became the established shape for all watering cans and has remained so, with little variation, ever since.

KNOW YOUR COMPOSTS

John Innes is a loamy recipe not a brand.

Nos 1, 2 and 3 contain the same ingredients, but the amount of food differs.

No. 1 is for pricking out seedlings.

No. 2 is for potting on seedlings.

No. 3 is for mature plants and for gross feeders like tomatoes.

Most composts contain only enough food to last for up to 6 weeks; after that the food runs out.

Use seed-sowing compost for seed.

Organic Tip ✔

Resist the urge to use Jeyes Fluid, bleach or any other chemical cleaner. These are damaging to the environment and will harm many of your overwintering friendly predators.

SECRETS OF SUCCESS

- Water is the undoing of most gardeners when it comes to seed-sowing. Don't over-water seedlings – put your index finger in the compost and sense how damp it is.
- Use the correct compost for seed-sowing. Mixtures are usually fluffy and light, but sifting the compost through your fingers adds more air.
- Water the compost before you sow.
- Always use mains water; water-butt water is less hygienic.

5 Sow Broad Beans under Glass

(late January)

RAISE YOUR broad bean plants under glass (in a cool greenhouse or cold frame) ready for planting out in March or early April. The cooler temperatures at this time of year encourage good root systems. You can sow two types – the taller, long-podded varieties and the shorter, hardier 'Aquadulce Claudia'.

Use modules (sectioned-off seed trays) that fit inside large seed trays: the 24 size (6 × 4) is ideal. Place the module into the seed tray and fill it with seed-sowing compost. Press one seed into each module and water the whole tray well. Cover with wire if you have a mouse problem – these large seeds are a lure!

Once the young beans reach 5cm (2in) in height, plant them out as soon as possible before the long radical root gets tangled up in the hole at the bottom of the module. Space them out, one plant every 22cm (9in) with 30cm (12in) between a double row. Put a series of canes round them and add some supporting string to prevent them flopping over other crops. Always top your canes with a cap or a small flowerpot to protect your eyes from damage when you pick. Decorative cane tops – mine are harvest mice on ears of wheat – are fun.

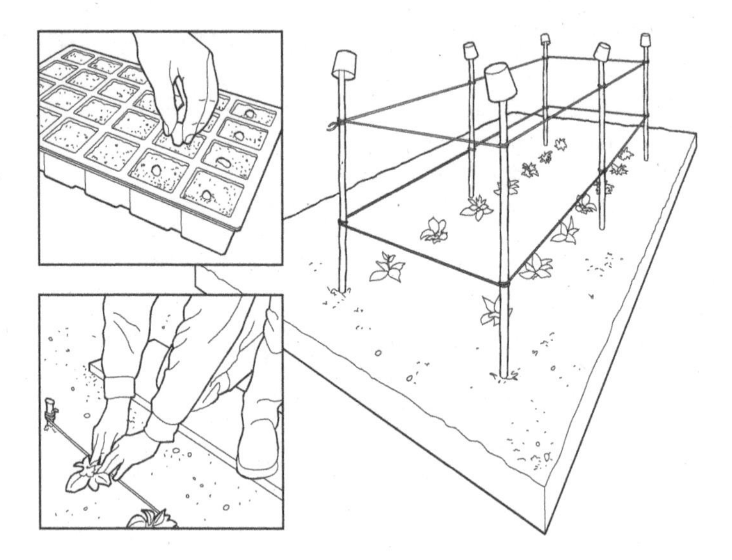

In March it's possible to sow the seeds straight into the ground. Use two seeds per hole to a depth of almost 5cm (2in). Plant a dozen extra seeds at each end of the rows for gapping up (i.e. filling any gaps in the rows). Cover with chicken wire if mice are a problem and then add canes.

Did you know? Broad beans originated in the region south of the Caspian Sea, but are now extinct in the wild. They were taken to America by the early Spanish settlers and became known as 'fava beans' from their Latin name. They didn't catch on in North America for a long time and were subsequently grown more in South America. Today Brazil is a leading exporter, but China is the world's biggest producer.

Organic Tip ✔

Broad beans inevitably attract blackfly because they emerge as these insects leave their winter host plants (Euonymus or spindle tree) to look for leafy bean plants. Pinch out and destroy any tips that get blackfly as soon as you see them.

Don't spray, even with organic mustard, garlic or soap sprays, as you will also kill helpful ladybirds and parasitic wasps.

SECRETS OF SUCCESS

- Mice and rats can devastate a row within an hour, even burrowing under snow. Keep them away by covering the seeds and plants with chicken wire.
- Once the pods are set well up the stem, pick out the tips so that the plants concentrate all their energies on filling each pod. You should expect 900g (2lb) of beans per 30cm (12in) of double row.
- Broad beans are self-fertile, but the yield is much higher when bumble bees cross-pollinate them. Wet or inclement weather can make a crop suffer because bees tend not to visit.
- Pick them regularly so that they keep producing flowers.
- Pick carefully, as broad-bean stems tend to be brittle. Use small scissors to snip off the pods if needed.

LONG-POD VARIETIES

'Jubilee Hysor' AGM
A prolific cropper and highly reliable broad bean with six to eight lime-green beans in each pod. Windsor bean flavour.

'Masterpiece Green Longpod' AGM
Slender pods filled with small green beans, but with a smaller yield than 'Imperial Green Longpod' and not such a strong flavour. Good for freezing.

'Imperial Green Longpod' AGM
Popular and reliable, with nine beans per pod. The beans have a good green colour, so are ideal for freezing.

'Meteor' AGM
An early crop with well-filled pods of pale-green beans.

6 Make a Runner Bean Trench

(late January)

IF DOUBLE digging seems a little onerous, it is well worth restricting your efforts to making a runner bean trench in readiness for May sowings or plantings. Starting now will give you 3 months to complete it. You can add all your soft organic matter, such as vegetable and fruit peelings, leafy weeds, spent cabbage leaves, plain paper and tea bags or tea leaves, etc. – all the things that would normally go on the compost heap. You can add comfrey leaves as well as pet bedding from rabbits and guinea pigs, but not cat or dog faeces as these are acidic.

Leave the trench open. The heap inside will begin to decompose a little, but once it is covered with soil (in March) the decomposition process will speed up as warmth is generated and moisture trapped below the soil's surface, helping your beans to get a better start than in ordinary soil. You can also make trenches like this for hungry crops like squashes, pumpkins and courgettes.

It is a good idea to buy runner bean seeds now before the best varieties sell out. You can raise beans in two ways. Either sow under glass in April and then plant outside – no earlier than June – or plant straight into the soil from mid-May through to July (see May, page 156).

Did you know? Recently bred red-flowered runner beans ('Polestar', for instance) have greater heat tolerance than heritage red-flowered varieties like Suttons' 'Prizewinner'. This means that 'Polestar' sets more effectively even if night-time temperatures exceed 16°C (60.8°F).

Organic Tip ✔

Slugs can nip out the growing points of runner beans and then they never recover. To lure predators away, plant young lettuces under and around your tripods or rows, then frisk them every evening at dusk when slugs are at their most active. Collect any you find and dispose of them.

SECRETS OF SUCCESS

- See page 158.

VARIETIES

For varieties, see May, page 159.

FEBRUARY

This is the month when the garden hovers between winter and spring and, although everyone longs for the moment when their jacket can be shrugged off with confidence, the experienced gardener secretly prays that winter keeps the upper hand until mid-March. A precocious spring, when the weather seesaws back and forth between warmth and chill, has particularly serious consequences for fruit growers. It's far better if the buds stay firmly closed in February and March.

Fruit trees, bushes and canes should look meticulously tidy, like well-trained soldiers standing to attention. Check supports and ties on trees and bushes, and replace as needed. If you're thinking of planting bare-root trees or bushes, they can still be planted now. If the weather is warm enough it's possible to plant containerized fruit into the ground.

On the vegetable front, the days are lengthening nicely and just occasionally a warm glimmer of sun manages to penetrate your clothing. That warmth goes straight through to the heart and promises better things to come. There's still a long way to go when it comes to ideal planting and sowing conditions, however, so watch and wait and try to catch the moment rather than sowing too early.

Certain crops are hardy enough to plant now, including garlic, onions and shallots. However, the temperatures will generally be too low for seeds to germinate in the ground. Our ancestors were adept at warming up the soil by covering it up with glass cloches and this can work very well. So if you're prepared to gamble, and early crops are always a little risky, you could cover part of your plot and sow or plant early.

Soil preparation and weeding are vital, so that once spring arrives everything is ready for the off. The winter frost should have broken down any rough digging by now, so all that is needed to create that elusive sowing surface, known as fine tilth, is a rake.

FRUIT

1 Cut Back Autumn-fruiting Raspberries

(early February)

AUTUMN raspberries crop from mid-August until early November on this year's new canes. This is the best time to cut the canes back. I prefer to reduce them all to ground level; however, some gardeners leave half the canes up and take half of them out so that they get June–July fruit on the uncut canes. This seems unnecessarily complicated – I prefer to grow summer-fruiting raspberries to pre-empt the autumn ones (see page 86). I also feel that cutting down completely now promotes stronger growth in spring, producing more vigorous canes.

Summer-fruiting raspberries can be tipped back now. Remove the top 15cm (6in) to encourage side shoots. They are properly pruned in summer (see page 227).

Collect up all the prunings carefully and either shred them or cut them up finely for the compost heap. Check all the ties and supports, making repairs as necessary. After pruning, dig lightly through the soil to disturb any overwintering raspberry beetles. If you have chickens, get them to help.

Did you know? Raspberries contain lots of vitamin C, plus other antioxidants, flavonoids and potassium. They are delicious eaten raw, they can be made into jam in less than 5 minutes (if the fruit is very fresh) and they also freeze better than any other soft fruit. Children love them.

Organic Tip ✔

Autumn-fruiting raspberries are the easier of the two to grow because their sturdy canes do not need staking as those of summer-fruiting varieties do. They are also less affected by raspberry beetle, which is more active when the summer varieties are fruiting. They do well in drier gardens as long as they are mulched because they fruit in cooler conditions – something raspberries enjoy. Water well in dry August weather.

SECRETS OF SUCCESS

- Raspberries tend to wander away from the row, so they need firm control. Chop out any unwanted canes in early spring just as they appear.
- Keep rows of raspberry canes weeded and give them an annual feed in late winter. Mulch with well-rotted manure.
- Don't thin within the row unless they are very crowded, and then only take out the weakest ones.

VARIETIES OF AUTUMN-FRUITING RASPBERRY

'Autumn Bliss' AGM
Firm, well-flavoured fruit that starts to ripen in August and carries on until late in the year.

'Polka'
The earliest of the autumn-fruiting raspberries. Very aromatic and high-yielding with a clean, fruity flavour. A possible replacement for 'Autumn Bliss'.

'Joan J'
A new, almost spine-free variety producing large red berries.

'Fallgold'
A yellow raspberry with a sweet flavour.

For varieties of summer-fruiting raspberry, see March, page 86.

2 Feed Your Strawberries

(early February)

AT THIS time of year strawberries can look very battered by the winter. If the weather is reasonable, tidy your plants lightly by cutting dead and damaged leaves away and then give them their first feed of the year, but try to wait for the worst of winter to pass. A potash-rich fertilizer will boost flower and fruit growth. Avoid nitrogen-rich feeds now; they will promote too much leafy growth. There are sprinkle-on granular feeds for strawberries, but liquid tomato food or comfrey tea (see page 326) work just as well.

As spring progresses, continue to tidy your plants and give the plot a weed. If flowers appear early, it's vital to prevent frost from blackening them. Cover them with hessian or horticultural fleece on cool nights.

Did you know? The strawberry we grow today is a descendant of an accidental eighteenth-century hybrid between a Chilean strawberry, *Fragaria chiloense*, and a North American species, *F. virginiana*. Antoine Nicolas Duchesne, botanist and gardener at the Palace of Versailles, grew both and, when planted side by side, they hybridized to produce large red fruit. The same hybrids were deliberately bred in England in the late eighteenth century by Thomas Andrew Knight, a horticulturalist and botanist who lived at Downton Castle in Herefordshire, and Michael Keen, a nurseryman of Isleworth in London, both of whom raised and named early varieties.

Organic Tip ✔

Strawberries are members of the rose family and, just like roses, suffer from 'rose-sickness' – a debilitating condition whereby roses planted in ground that has recently held other roses do not thrive. If you want to replace a wintertime fatality in an existing strawberry bed you will need to replace the old soil with fresh soil in which strawberries have not been grown.

SECRETS OF SUCCESS

- There are three types of strawberry plant: cold-stored plants are available between March and June; misted-tip plants (grown from cuttings) can be bought in August (see page 230); and freshly dug plants are available during October. Plant all as soon as they arrive.
- You must conduct a war on three fronts in order to grow strawberries successfully: against disease, slugs and birds. Tidiness and regular care are important.
- Cut away dead and damaged leaves close to the crowns. Do not try to pull them off – you will damage the crown.
- If you have missed any runners, remove them now.
- Do not add bulky organic matter as a substitute for liquid feeds – it encourages slugs.
- Rotate your crop and discard plants once they are 4–5 years old. Avoid planting where potatoes have grown before or they will be at risk from verticillium wilt. Do not plant potatoes close to a strawberry bed for the same reason.

VARIETIES

'Honeoye' AGM (early)
A good flavour with conical, orange-red, glossy berries. Good in cool districts.

'Sonata' (mid-season)
Large, sweet fruit that resists extreme weather, including very hot spells and heavy rain. Resistant to powdery mildew.

'Hapil' AGM (mid-season)
Heavy crops of large, bright-red, glossy berries with an excellent sweet flavour. Crops especially well on light soils and in drier conditions.

'Fenella' (late)
Heavy-yielding with large, glossy berries that have an aromatic, sweet flavour. Good resistance to verticillium wilt and crown rot, and the fruit can tolerate very heavy rain.

'Florence' AGM
Disease-resistant, heavy-cropping strawberry with large, dark red fruits with a very good flavour. Shows some resistance to vine weevil.

For further varieties of strawberry, see March, page 72.

3 Plant and Train Blackberries

(mid-February)

IF YOU haven't grown a cultivated blackberry, it's well worth doing so. The flowers appear after the frosts have stopped, so they make good productive plants for cold gardens or those in frost pockets. They can be trained in a corner of the garden as long as they are given the support of a wall or fence and allowed to climb into a brighter position. They are not fussy about soil, even seeming to tolerate heavy soil. Given a reasonable position, they can produce over 5kg (10lb) of fruit per plant.

Choose the variety to fit the space available. Some need as little as 2m (6ft); others, such as 'Fantasia', need 4.5m (15ft). Hybrid varieties are generally smaller. Blackberries can be planted any time in the dormant season, November–March. Dig in 5–8cm (2–3in) of well-rotted compost or manure thoroughly at the planting site.

The key to growing cultivated blackberries successfully lies in pruning and training. Blackberries fruit on last year's canes from late summer until the first frosts. Next year's canes sprout from the rootstock from midsummer on and their vigorous growth gets in the way of picking if they are not contained.

The answer is to train the new canes away from the old canes. The easiest way is to train all canes one way on the supporting wire frame one year and the other way the next. This one-way system is wasteful of space, however, since only half the frame is productive in any one year.

A second way is to train the new canes vertically, having already trained last year's canes well to the left and right to leave a gap for them. The new canes are gradually gathered together into a tight bundle which is fixed to the frame. The old canes are cut away after fruiting and the bundle of new canes is left in place over winter.

Now is the time to train the new canes into their fruiting positions. Undo the bundle and tie in the flexible canes horizontally to the frame. If the canes are long they can be woven up and down.

Dig over the soil underneath the canes to uncover any pests. With luck the birds will eat any overwintering raspberry beetles; these also attack blackberries. Then top-dress the area with a general fertilizer.

Did you know? In medieval England it was widely believed that the devil spat or urinated on all the blackberries in the hedgerows on Michaelmas Day, 29 September, so country-dwellers eschewed them after that date. Late fruit does tend to ferment, giving it a strange flavour.

Organic Tip ✓

Blackberries like warm sunshine and they grow best along the Pacific coast of Oregon. Try to train yours into positions where the branches pick up afternoon sunshine even if the roots are in shade.

SECRETS OF SUCCESS

- Space plants in rows 1.8m (6ft) apart if growing more than one.
- Erect wire frames to support the branches and make picking easier. These should be in place prior to planting.
- Train the canes horizontally: they will produce much more fruit.
- Cut out the old fruiting canes after flowering to promote vigour.
- Keep the area around blackberries well weeded.

VARIETIES

'Bedford Giant'
An early-fruiting variety producing large clusters of flavourful, round berries from late July to August. A very vigorous variety – not suitable for a small garden.

'Loch Ness'
This thornless blackberry produces very high yields of top-quality fruit. Berries are large, very firm and glossy black. The most successful commercial variety in Britain.

'Karaka Black'
A very early variety from New Zealand. This 'King' blackberry produces elongated fruit by July and carries on for 2 months. The fruit is easy to pick.

'Oregon'
A thornless blackberry with divided leaves that crops in late August or September. Can be safely planted against a wall or fence as it has smooth stems. A lighter cropper.

4 Prune Newly Planted Plums

(mid-February)

PLUMS MAKE excellent fruit trees for small gardens because many varieties are self-fertile – so you can plant one tree and get a crop. The golden rule is that you never prune plums in winter because of the risk of silver leaf disease. The time to prune established trees throughout their productive life is in the second half of July (see page 208). However, newly planted plum trees will need pruning, just as the buds are breaking, for the first 2 years.

In the first year cut the main leader (the branch forming the apex of the tree) back to 1.5m (5ft). Then shorten the laterals by half. Any branches that are too close to the ground should be removed at their source. Prune again in late July – then in the following March the only pruning that is required is to tip back the main leader by two-thirds.

Did you know? Plums have been grown in the Vale of Evesham for hundreds of years. One authority, Ron Sidwell (1909–93), who spent his life studying their requirements and eventually became vice principal of Pershore College, was famous for being able to name every plum variety from its stone alone. He mapped the weather all over the region to discover the most frost-free area for commercial plum-growing, finally identifying a hamlet called Little Paris on Bredon Hill as the most frost-free spot. He moved there and established a large garden, Bredon Springs, containing many frost-tender plants from the southern hemisphere. Sidwell went on to breed the famous Bird Series of penstemons, including 'Raven', 'Blackbird' and 'Osprey', resurrecting the penstemon's popularity once again in the 1960s.

Organic Tip ✓

The fungus that causes silver leaf is closely related to wood-decomposing fungi. Like them, it fruits on wood from late summer on and can survive for a long time in cut wood. So infected branches can be cut away during the summer without any risk of spreading the disease, but all wood and leaves should be burned to kill the fungus.

SECRETS OF SUCCESS WITH PLUMS

- Plums (like all stone fruits) flower early in the year and need to be planted in areas that escape spring frosts. They also need a sheltered position to encourage pollinators. Gages need a warm position and can be difficult to grow in cold districts.
- Plums prefer well-drained, moisture-retentive soil and do best in areas where summer rainfall tends to be plentiful.
- Mulch with well-rotted manure in mid-spring to preserve moisture and increase nitrogen (see page 20). Top-dress with potash-rich fertilizer in late winter.
- Plum trees can be wayward in shape, so branches may need tying down, raising up or trimming back during the growing season.
- Branches can snap under the weight of fruit, so thin out the crop if this looks likely.
- Pick some plums with the stalk attached a little before they become fully ripe: this way they will keep for 3 weeks in a cool place.
- Remove any mummified plums in the autumn – these will harbour brown rot spores over winter and cause re-infection the following year.

VARIETIES

'Opal' AGM
A new, extra-early variety with medium-sized, yellow-fleshed, red fruits. Very good to eat, but also makes good jam. Pick in late July. Self-fertile. Pollination Group C.

'Czar' AGM
A dark, purple-black plum for cooking and jam but it can also be eaten raw when fully ripe. Yellow-green flesh with a good acid flavour. Pick in early August. Reliable and seems to shrug off frost. Self-fertile. Pollination Group C.

'Marjorie's Seedling' AGM
The best late plum for cooking and eating, although it forms a large tree. It flowers later than most and the blossom appears after the leaves (which is unusual), so it misses the early frosts. Pick in late September. Self-fertile. Pollination Group B.

'Victoria' AGM
The most popular plum with both amateur and commercial growers because it crops reliably year after year. Yellowish-green flesh and a pink-bloomed skin, but the wood is brittle. Pick in mid-August. Self-fertile. Pollination Group C.

5 Prune Cobnuts and Filberts

(late February)

BY LATE February the hazel catkins will be sending up clouds of yellow pollen. This is the traditional time to prune the trees, so that the magic yellow dust (the male pollen) falls on to the small red female flowers, which resemble tiny sea anemones.

The ideal nut tree should be a multi-stemmed bush, roughly 1.5–2m (5–6ft) in height for easy pruning and picking. Prune between January and March, aiming to create a bowl shape with an open centre containing eight or so branches radiating from a central stem. Pruning is simple: you saw some of the older, thicker branches out at the base. If this is done now, while they are devoid of leaf, these strong pieces of wood can be used as bean poles or pea sticks, depending on their size. Growing hazels is a sustainable system, providing both nuts and staking material.

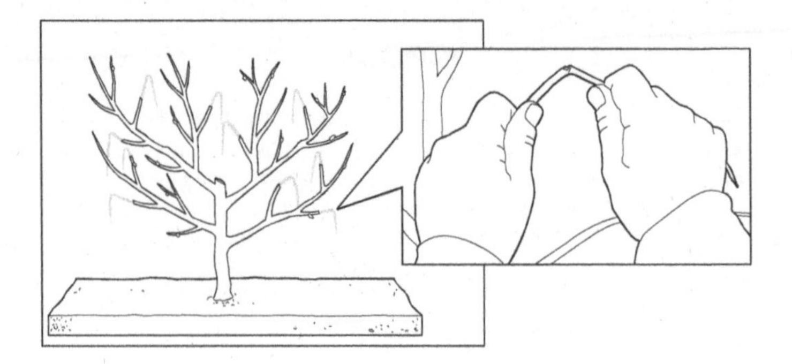

In late summer the newer shoots will look like narrow whips. Bend the upper half over (leaving it attached) to slow the sap. This will encourage more flower buds. In autumn the warm-brown nuts will nestle in a lighter green ruff. The time to harvest is when those ruffs turn yellow. Unfortunately the squirrels often beat the gardener to the crop.

Nutteries can be very decorative because they make an ideal canopy for all manner of woodlanders including hellebores, snowdrops and spring bulbs. So your nut trees can be incorporated into an ornamental garden – unless you have room for a nuttery like Sissinghurst's. If you do, so much the better – hazel wood is the best for staking plants.

Organic Tip ✓

Dig and weed the ground under nut trees during winter. This should expose the larvae of the nut weevil, a troublesome pest, to the winter frosts.

Did you know? The name 'hazel' comes from the Anglo-Saxon word *haesel* meaning a head-dress or bonnet, referring to the shape of the outer husk. Cobnuts and filberts come from different species of hazel. Cobnuts (*Corylus avellana*) are bred from our native hazel and produce round nuts with a short husk. Filberts come from a south-eastern European species called *Corylus maxima* and the nuts are longer, with a pointed top and long husks. They take their common name from St Philbert, whose feast day is 20 August, when picking starts.

SECRETS OF SUCCESS WITH HAZELNUTS

- Plant in a sunny, open position where the hazel can expand as a specimen tree. Leave 3m (10ft) between trees if possible. Keep young trees well watered in their first three growing seasons.
- Cobnuts dislike heavy, waterlogged clay. They do well in poor, well-drained sandy loams, such as parts of Kent, where 'plats' (cobnut orchards) have always been commercially successful. A soil that is too fertile will tend to produce vigorous trees that don't crop well.
- Young branches should be trained to near horizontal (a hoop can be used) and the lateral spread is encouraged by always pruning to just above an outward-facing bud.
- Propagate cobnuts by bending established suckers over and pegging them down.

VARIETIES

'Webb's Prize Cobb'
A new, reliable variety that produces bigger clusters of large, long nuts than 'Kentish Cobb' (see right). A good choice for northern gardens.

'Hall's Giant Cobnut'
This French variety is commercially grown in its native land. It is hardy and vigorous and produces a heavy crop of very large cobnuts. Must have cross-pollination to produce the best crops.

'Kentish Cobb'
A classic, heavy-cropping variety grown in Kent – probably the most commonly grown. Despite its name, it is a filbert.

'Pearson's Prolific' (syn. 'Nottingham Prolific')
Superb, compact habit makes this cobnut very suitable for the smaller garden. Good crops of medium-sized, round nuts. Good pollinator of other varieties.

6 Prepare Fruit-cage Netting and Fleece

(late February)

FEBRUARY is a quiet month in the fruit garden once winter-pruning has finished, so it is an ideal time to check over the fruit-cage netting. If you took the top cover off to prevent snow from bending the uprights, reinstate it now. Take at least one side panel off to allow larger bees to access the fruit flowers, or choose a larger mesh to allow them in. Although fruit is often self-fertile, crops are normally higher if plants are cross-pollinated.

Strawberries may flower as early as April in warm springs, but they are very susceptible to frost. Be prepared to fleece the flowers on cold evenings with hessian or heavyweight horticultural fleece. Use bricks or stone as weights if it is likely to be windy. Peaches and apricots will need fleecing too.

Did you know? Nets have been used to protect fruit since Roman times at least. The writer and farmer Columella (AD 4–70) describes nets made of broom being used to prevent birds from eating pomegranates. The diarist John Evelyn describes using nets in 1686, but the earliest wire fruit cage was a nineteenth-century invention designed to prevent birds from eating cherries.

Organic Tip ✔

If you do fleece strawberries and other fruit, remember to remove the blanket every morning to allow pollinators access to the flowers.

SECRETS OF SUCCESS WITH NETTING

- Pick the right mesh size for the job.
- Treat mesh gently – undesirables will find even the smallest hole.
- If you can, add a small (10–15cm/4–6in) protective wooden plinth around the foot of the fruit cage. A plinth made from rustic planking – 'edged slabwood' in the trade – looks good too.
- Take the roof netting off in winter. Many fruit cages are destroyed by the weight of snow.

VEGETABLE

1 Sow Leeks under Cover

(early February)

LEEK SEEDS can be sown under glass now. This hardy vegetable prefers to germinate in cool conditions and once the seedlings are up you don't have to worry about cold nights. However, all members of the allium or onion family are shallow-rooted. They cannot seek out moisture from the depths, so the seedlings can dry out easily. If young leeks become water-stressed they tend to bolt (run to seed), so it's vital to keep your seedlings damp. For this reason I always try to keep them out of direct sunlight in the greenhouse. Leeks can be sown outside straight into the ground if you wish, but not until mid-March at the earliest.

The easiest way to raise leeks of the right size under glass is to use 6 × 4 modules. Place two seeds in each one, then weed out the weaker seedling if necessary, leaving the other to fill the space. Once the young plants are pencil-thick (after 10 weeks or so) they are ready to go outside. Make deep holes with a dibber, drop one leek plant in each and then fill the holes with water. The tops and roots do not need trimming. Each hole needs to be 15cm (6in) deep with 22–30cm (9–12in) between each. Rows should be between 30–38cm (12–15in) apart. Wider spacings aid air flow, helping to prevent diseases like rust.

Transplanting usually takes place in the second half of May, although leeks can be planted out up until mid-July. Most of the growing takes place in autumn and leeks are invaluable – they almost always come through the hardest winters unscathed.

Did you know? The Ancient Egyptians grew leeks in the time of the pharaohs. The Greeks and Romans also loved them. Nero was said to eat them with olive oil to preserve his voice. The leek has a powerful reputation as a medicinal plant and it was also thought to have magical properties. If you wore a leek it would protect you from battle wounds, stop you from being struck by lightning and keep away evil spirits.

Organic Tip ✔

Leeks do not do well on compacted soil, so adding some well-rotted organic matter really helps this crop dramatically. Keep the weeds down - but remember those shallow roots.

SECRETS OF SUCCESS

- Fork over the soil a day or two before planting; this makes it easier to get the dibber in.
- Choose a good F1 variety for early sowing – the seeds germinate better.
- Once the seedlings have been 'dibbed in', water the whole plot well and keep it damp. Don't dribble on water; use a sprinkler if possible and give the whole area a thorough soaking for at least 2 hours in the latter half of the day.
- Continue to water in this way whenever dry weather occurs in the first month after planting. Once they look established, leave them to their own devices.
- Lift leeks as you need them, using a fork, as they are best eaten very fresh. They will not store for long: they become tough.
- If a leek bolts, snap off the flowering stem.
- You can earth leeks up to blanch more stem.

VARIETIES

'Oarsman' AGM
A smooth-skinned, maincrop F1 hybrid leek that is much kinder on the stomach than the thickly textured, cellulose-packed 'Musselburgh'. Long, sleek shanks that cook sweetly.

'Carlton' AGM
An earlier F1 variety producing mid- to dark-green flags, but this variety bolts more easily than some so is not for dry gardens.

'King Richard' AGM
An early, pale-green variety that resists bolting well.

'Apollo' AGM
Vigorous plants with attractive dark-green leaves that fan out from a thick white shank. Ready from mid–late winter and resistant to rust.

2 Plant Garlic

(early February)

GARLIC IS another shallow-rooted member of the allium family that needs lots of water to crop well. In the wild this plant from the high mountains is woken up by melting snow and warmer temperatures. Hot, dry weather triggers dormancy and, as a gardener, you should be harvesting once the leaves begin to wither and yellow.

There are two types of garlic – softneck and hardneck. The softnecks produce plump, white bulbs and, if planted now, they can be harvested from mid-July and will keep until April. Softnecks tend to bulb up as the days shorten after the summer solstice. Hardneck cultivars produce an edible flowering stalk (or rocambole) and this must be snapped off as soon as it emerges to encourage the bulb to swell. Their flavour is stronger and the bulbs are usually more colourful. Hardnecks are planted in September and October and harvested in June. They generally keep until January.

The planting technique is the same for both. Break the bulb into cloves just before planting and place the individual cloves 2.5–5cm

(1–2in) below the soil surface, roughly 15cm (6in) apart. Lift and harvest as soon as the leaves begin to yellow.

Did you know? Egyptian slaves were given a daily ration of garlic to ward off illness and increase strength and endurance. During the reign of Tutankhamen 7kg (15lb) of garlic would buy a healthy slave.

SECRETS OF SUCCESS

- A sunny site is vital for garlic.
- Don't try to grow supermarket garlic: invest in proper varieties.
- Snap off any flowering stalks if they appear on hardneck varieties: this will fatten the bulbs.
- Water is vital for large, plump bulbs.
- Dry the bulbs thoroughly before storing or plaiting.

SOFTNECK VARIETIES

'Venetian Wight' (from the Po Valley in Italy)
Small, hard, white garlic that will keep until spring. Intense flavour and quite hot.

'Provence Wight' (from the Drôme Valley in Provence)
Grows well in Britain and the fat, juicy cloves are perfect for adding some Mediterranean flavour to vegetable and fish dishes. Can also be planted in autumn.

'Solent Wight' (from the Auvergne in France)
The most robust garlic in overall terms of eating and keeping. Large, dense white bulbs with a good flavour. The easiest garlic to plait.

'Tuscany Wight' (from Italy)
Large, fat cloves all the way round the bulb. Classic Italian flavour, good with chicken and lentils.

For more garlic varieties, see August, page 240.

Organic Tip ✔

Always lift garlic as soon as you spot the leaves fading or yellowing, otherwise the bulbs will sprout again and will be no good for storing.

3 Cloche Areas Ready for Sowing

(*mid-February*)

OUR ANCESTORS were far more inventive than we are when it came to growing vegetable and fruit crops, and they used protective glass in a variety of ways to produce precocious crops. We, however, have a whole range of products that weren't available to them. They include child-friendly plastic, horticultural fleece, Ecomesh and all sorts of netting.

Now that the February sun is beginning to gain some warmth, it's possible to cloche some areas where early crops will be sown. These might include carrots, parsnips, beetroot and spinach. Lightweight polythene cloches are not suitable for windy gardens – they do a disappearing act. Even those with holes that allow the wind to pass through tend to end up decorating the local hedgerows.

However, the heavier plastic cloches are useful for getting areas of ground ready for sowing. Prepare the ground well so that the soil is fine, then water it well if it is dry. Cover with your cloche, making sure the ends are blocked off. Many a cat has adopted the warm ground underneath. Open the ends regularly to allow the air to circulate. After 2 weeks sow the seeds, water again if necessary, and cover once again with the cloche. If cold weather is forecast, fleece over the plastic with a double layer.

Did you know? The giant glass belljar was introduced into Britain in the early seventeenth century and was immediately known by its French name of *cloche*. The 'big glass hats', as they were called, were covering the melon beds of England by 1629. By 1677 English-made square glazed less-breakable covers with lead frames and lids had taken over.

Organic Tip ✔

When you plant out your first courgette, squash, cucumber and pumpkin plants at the beginning of June, cloching them at night (with a sturdy plastic bell cloche or similar) produces plants double the size of the uncloched ones.

SECRETS OF SUCCESS

- Invest in sturdy materials that will last for a few seasons. Light polythene and wafer-thin plastic are not rugged enough.
- Ventilation is the key to success when covering plants, because still, damp air encourages fungal diseases like damping off and botrytis. Make time to uncover your seedlings or young plants every day, exposing them to fresh air.
- Water regularly, because rain cannot penetrate most barriers and this includes Enviromesh and horticultural fleece.
- Use wire or wood supports to keep any coverings away from the foliage. Canes can be used.
- Early strawberries can be protected with a makeshift tent covered with thick polythene.

TYPES OF CLOCHE

Glass is still the best material for cloches. It lets in the most light and insulates the plants, but it isn't child-friendly.

Thick plastic bell cloches have a ventilator at the top and are heavy enough not to blow away on windy days.

A double layer of horticultural fleece is the most effective protection on cold nights. This light material will last for many years, but it does absorb water and it will weigh plants down on wet days.

4 Force and Divide Rhubarb

(mid-February)

OVER THE years mature rhubarb clumps can become very large and this is the best time of year to divide them, just as the large, dome-like buds are breaking dormancy. Lift the whole crown and, using a spade, split it into chunks containing four or five buds. Replant in enriched soil containing garden compost, making sure that the top of the clump is just above the ground. Do not pick any stems in the first year or two and always remove any flowering spikes.

This is also the best time to plant new rhubarb crowns because the soil is warm enough to promote growth. Choose a site that is well-drained, fertile, and preferably in full sunlight because rhubarb grows naturally in open sites on moist riverbanks. It pays to prepare the soil well by eradicating all perennial weeds and digging down deeply. Add some garden compost or loam-based compost, such as John Innes, to the bottom of the hole because this will retain moisture. Rhubarb will not grow well in water-logged positions, but you can create a raised mound to aid drainage. The tip of the crown should be at soil level and visible. This is important because cold weather can penetrate the crown and signal a period of dormancy. You should also clear the foliage in late autumn so warmth can trigger growth in the following spring. Don't pull any stems for the first year or two.

Did you know? *Rha* was the name the Greeks gave to the River Volga and to the edible plant that grew on its banks. Later, the plant became known in Latin as *rha barbum* ('barbarian' or 'foreign') and from this the name 'rhubarb' is derived. The first forced rhubarb was an accident. Workmen at the Chelsea Physic Garden in 1815 covered the rhubarb patch with builders' rubble. When it was cleared, long, pale stems were uncovered. Commercial forcing in Yorkshire's 'Rhubarb Triangle' began in the 1880s and many of the rhubarb varieties we grow today date from this golden era. The Triangle once covered a 30-square-mile area between Leeds, Wakefield and Bradford, but now it is reduced to a 9-square-mile area between Wakefield, Morley and Rothwell.

SECRETS OF SUCCESS

- Choose an open, sunny site and prepare the soil by working in plenty of farmyard manure or compost before planting.
- Plant in spring where possible, placing new crowns 1m (3ft) apart with the buds just above the surface. Don't pull any stems until the second year of growth.
- Never cut rhubarb. The technique is to pull and then twist very gently from the lower stem.
- Stop harvesting at the end of May to allow your plants to recover.
- If a stressed plant should run to seed, remove the flowering spike straight away. Water, feed and mulch lightly.
- Tidy rhubarb in autumn so that slugs do not have anywhere to hide.

Organic Tip ✔

Rhubarb prefers damp summer weather and a dry winter. Mulching (with well-rotted compost or straw) will help to achieve both, but do not cover the crowns deeply.

VARIETIES

'Timperley Early' AGM (early)
So early it's probably better not to force it. The slender, long, pink-red stems have a tart flavour that makes this an excellent crumble filler. Not a prolific cropper – but a must for all rhubarb lovers.

'Queen Victoria' (mid–late season)
Colourful, strong red stems, easy and prolific. This heritage variety still holds its own today. Vigorous, making huge clumps, so perhaps not for smaller gardens.

'Hawke's Champagne' AGM (early–mid season)
Delicately thin, long scarlet stems with a sweet flavour from early spring. An old variety, but easy to grow and ideal for forcing. Attractive appearance.

'Raspberry Red' (mid–late season)
An old Dutch variety, recently reintroduced, with sweeter red stems. Given a sunny, open position it is a heavy cropper.

5 Plant Shallots

(late February)

TRADITIONAL garden wisdom decrees that shallots should be planted on the shortest day and harvested on the longest because they take 24–26 weeks to mature. However, we suffer from wetter, warmer winters than we used to have and I've found that the results are much better from late-February plantings. Plant now and you can harvest in August when the weather is at its warmest.

The flavour of shallots is subtler and more aromatic than that of onions, so they are well worth growing, especially for the keen cook.

However, despite being smaller in size, they actually take up more space in the garden than onions. The foliage splays outwards as the cluster of shallots forms and you should get between seven and nine babies per set. Choose a sunny position and plant them 22cm (9in) apart in rows 22cm (9in) apart. Leave the upper third of the bulb showing because shallots are prone to rotting in damp soil.

Shallots grow better and ripen better in sunnier summers and, although they are not quite as hardy as onions, they often store for longer. There are red and yellow varieties, but the yellow ones are the easiest to grow and they also store for longer. The key to good shallots is watering in the early stages and also putting them on fertile ground. Ideally, the onion plot should have been manured or enriched in the months leading up to planting.

Did you know? The botanical name for the shallot, *Allium ascalonicum*, is derived from Ascalon, a place in Palestine where shallots are thought to have originated. The Crusaders possibly brought them to England in the twelfth century, but the Greek writer Theophrastus (371–287 BC) referred to them in his writings, as did Pliny the Elder in the first century AD.

VARIETIES

'Golden Gourmet' AGM
The heaviest-cropping, golden, ball-shaped shallot, producing substantial bulbs.

'Pikant' AGM
A Dutch variety. The best red, with lots of layers of brown-red skin and a very rounded shape. Good flavour.

'Longor' AGM
A long, slender 'Jersey long' shallot with golden skin, almost pear-shaped, with a strong flavour.

'Jermor' AGM
A copper-coloured, long shallot, widely grown commercially in Brittany. Pink-tinted flesh and a good flavour.

Organic Tip ✔

Dry springs are the undoing of shallots because they need to establish roots in order to multiply successfully. Water them in the early stages of growth whenever it's dry.

SECRETS OF SUCCESS

- Find a garden hotspot where the sun shines and be prepared to water them well in warm, dry weather.
- Source good-quality, plump sets that feel firm to the touch and that have no mould.
- Avoid planting on newly manured ground – it scorches the roots.
- If you find shallots difficult to grow, opt for globular, golden varieties – they are the easiest to grow and they store for longer.
- The banana-shaped shallots and the red shallots do not store for as long as the yellow. If you grow them, use them up before the end of December.
- Lift the shallots away from the ground with a fork in August to help the drying-out process, gently teasing the clusters of bulbs apart.

6 Plant Onion Sets
(*late February*)

IT'S ALSO time to plant onion sets, and the first job is to trim off the little wispy tops with a small pair of scissors; otherwise the birds will tug them out. They should take 20 weeks to reach maturity. Later in the year (usually by mid-March), there are more-expensive, heat-treated sets for sale. These are planted in March and April and their biggest advantage is that they rarely (if ever) bolt.

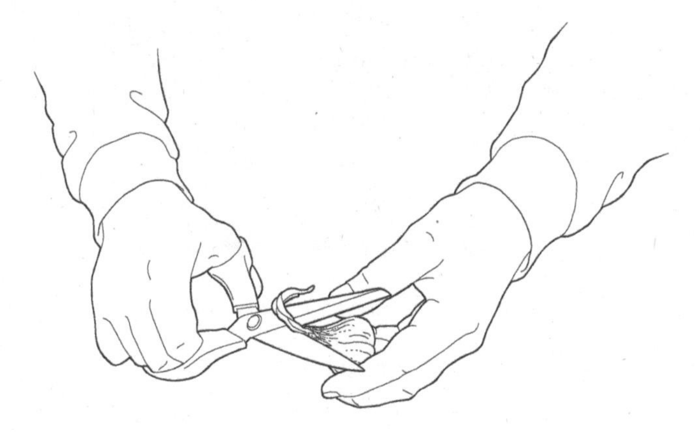

Bolting is more of a problem in areas that get dry springs and searing winds. If that describes your conditions and you've experienced lots of bolting onions, use the heat-treated sets. If bolting does occur, remove the flower bud and stem immediately.

Space your sets 15cm (6in) in rows 22cm (9in) apart. Push each set into the ground so that the tip is at ground level and use lines to keep the rows straight. It's essential to keep down the weeds with a small onion hoe, because, like shallots, onions are shallow-rooted and can't cope with competition. Having neatly aligned rows is a huge help when hoeing.

Did you know? There is no wild equivalent of the onion, so it must have been domesticated for thousands of years. Carvings appear on Egyptian tombs that are 5,000 years old. When King Rameses IV died, onions were placed in the eye sockets of his mummified body as a symbol of eternal life. The Romans are credited with bringing the onion to Britain.

Organic Tip ✔

Red-skinned onions should be planted 3 weeks after the golden-skinned varieties because they prefer warmer conditions. Generally, red-skinned onions and shallots do not store for as long as golden-skinned varieties.

SECRETS OF SUCCESS

- Water well in the early stages whenever the soil is dry to develop the root system.
- Keep onion beds well weeded by hoeing them regularly with a small hoe. This will create a fine layer on the top which mulches the soil beneath and keeps moisture in.
- Plant onions in a bright position.
- Do not bend the necks artificially.
- In August you should be able to lift the bulbs away from the soil with a fork. A couple of weeks afterwards it's possible to lay the onions on their sides.
- Laying onion bulbs on a simple framework will dry them for storage.

VARIETIES

'Sturon' AGM
A globe-shaped, yellow-brown onion with juicy flesh. Produces rounded, medium-sized bulbs quickly and rarely bolts.

'Red Baron' AGM
A dark, glossy, red-skinned onion with sweet flesh and a good flavour. Still the best red variety.

'Centurion' AGM
A golden onion with a flattened shape and a distinctive pale skin the colour of straw. A heavy yielder with a crisp flavour.

'Santero'
A new, downy, mildew-resistant F1 onion with coppery-brown skin. Good in the drier parts of the country where mildew is prevalent.

SPRING TASKS

FRUIT

- Weed round fruit trees and bushes.

- Be prepared to fleece early blossom on almonds, peaches, apricots and nectarines, but make sure the pollinators can still reach the flowers. Early plum and pear blossom can be vulnerable to late frost.

- If you have strawberry plants under glass you may need to pollinate the flowers. Use your hands and fingers to distribute the pollen.

- Prune blueberries and apply an ericaceous fertilizer.

- Erect codling moth traps.

- Check blackberries and loganberries and make sure their canes are spread out and attached to wires.

VEGETABLE

To Do

- Prepare soil for planting
- Fertilize with blood, fish and bone (see page 317) or powdered or pelleted chicken manure
- Keep weeding
- Water on slug nematodes in key areas
- Put up bean canes
- Protect early sowings of potatoes by earthing them up (i.e. mounding soil up around plants, leaving just the tops showing) or covering them with fleece
- Harden off tender plants before planting
- Fleece vulnerable plants on cold nights
- Remove and clear all winter vegetables
- Uncover rhubarb crowns that have been forced

Sow under Glass

- Leeks
- Sweet and chilli peppers
- Tomatoes
- Aubergines
- Cucurbits – cucumber, squash, pumpkin, courgette, etc.
- Lettuce
- Globe artichokes
- Brussels sprouts, purple sprouting broccoli, kale, cabbages
- Runner and French beans
- Sweetcorn
- Celery
- Chicory and endive
- Herbs

Sow Outdoors

- Beetroot
- Carrot
- Spinach
- Broad beans
- Peas
- Parsnips
- Turnips
- Swede

Plant

- Asparagus crowns
- Potatoes
- Rhubarb
- Broad beans grown in modules

MARCH

In theory March is the month when spring arrives, but the weather is never predictable, so this is a month for watching and waiting for the correct growing conditions. The grass should be growing, along with the weeds, and the birds should be nesting before you start outdoors. If you sow too early, when the soil is still cold and wet, most seeds just sit there and fail, so bide your time and go with the weather rather than the calendar. This may be frustrating, but years of experience have taught me the wisdom behind the proverb 'patience is a virtue' when it comes to growing vegetables. Later-sown crops will catch up quickly.

March days can be warm, but night-time temperatures often plummet and there are still several weeks when frost could strike, so crops planted in March have to be totally hardy. Tender crops (including tomatoes, peppers, cucurbits and aubergines) are best left until later in the season, because even a slight frost will kill them. Their growth can be permanently checked by cold nights and generally plants don't recover. You can start seeds and young plants off in a warm greenhouse in March, but don't put any tender crops outside until late May or early June.

Woody fruit crops do well planted in March and, although the bare-root season is over by now, there will be lots of container-grown fruit bushes and trees for sale. Use a specialist supplier and prepare the soil well straight after ordering, so that you can plant them as soon as they arrive. Double dig the area, if you can, and add organic material such as well-rotted manure, or home-made garden compost.

This is also a good time to get rid of fruit that isn't performing. Strawberries, for instance, lose vigour after four to five years and need to be replaced by new virus-free stock. They also need to be planted in a different part of the garden. If you're removing any fruit, do not replant the same crop in the same spot. Give it a fresh site, as the soil may either harbour diseases and pests, or be exhausted.

FRUIT

1 Make a New Strawberry Bed

(early March)

MOST PLANTINGS of fruit are semi-permanent: once in position they stay there for decades. Strawberries, however, are moveable feasts that inhabit the same spot for 5 years at most; after that their productivity plummets and they need replacing. For this reason they are often best incorporated in the vegetable garden, although they must never rub shoulders with potatoes or tomatoes as they share a disease – verticillium wilt.

Strawberries are traditionally planted 37–45cm (15–18in) apart in rows 90cm (3ft) apart. This allows enough space for hoeing between plants. It also allows you to water without damaging and splashing the fruit. They can, though, be grown very successfully in small (2.4 × 1.2m/8 × 4ft) beds. The plants and rows are spaced more closely, at 30cm (12in). Eighteen plants in such a bed will give you 13kg (30lb) of fruit over a 5-week period.

Choose a warm, sunny position and try to give your plants good drainage and good soil. Avoid a frost pocket; the crop will be ruined if frost catches the flowers. Before planting, incorporate plenty of bulky organic matter. Two barrowloads on a 2.4 × 1.2m (8 × 4ft) bed will be ample. Be sure to pick out any perennial weeds, as it is impossible to remove them once the bed is established.

Take care when planting. Trim back the roots to roughly 10cm (4in), then spread them out in the hole. Ensure that the base of the crown rests lightly on the surface. Planting at the correct depth is important: if the crown is planted too deeply it will rot; if it is planted too shallowly the plants will dry out and die.

It is often a good idea to begin a new strawberry bed whilst the old one is 3 or 4 years old – so that there isn't a gap in production. When making a new bed it's best to order in fresh plants that are guaranteed virus-free. Strawberries are very susceptible to virus and you must always buy certified virus-free plants: do not accept next door's freebies, however tempting they may be. Breeders constantly launch new varieties, so be adaptable about which to grow. The home gardener is always advised to scrap their plants every 4–5 years and start again. This makes sound sense.

There are three types of plants on offer in the nursery trade. The conventional plants raised from runners are sent out in early autumn. These will flower in the following summer, but all flowers formed this year should be removed to allow the plant to concentrate on producing a good root system. This will increase the crop in the third year hugely.

Recently two more types have appeared. 'Frozen' plants (which have been cold-stored) are available between March and June and arrive as bare-root bundles. They should be planted as quickly as possible and kept well watered during their first season. They should fruit within 60 days of being planted.

Misted-tip plants (grown from cuttings) are available in August – see page 230.

Did you know? Most strawberries need short days to form their flower buds and they use a pigment called phytochrome to tell them whether it's day or night.

There are two types. Short-day varieties crop in June and July. Perpetual varieties ('everbearers') crop in flushes throughout the summer, but these do not have the traditional sweet strawberry flavour. Instead they have the aromatic tang of alpine strawberries. Perpetual-fruiting varieties are normally grown for 2 years only and they don't usually produce runners.

Organic Tip ✔

Incorporate organic material into the soil thoroughly when forming a new strawberry bed. Any left lying on the surface will only encourage slugs. A double-dug bed is best of all (see page 308).

SECRETS OF SUCCESS

- The earlier in the year you plant strawberries, the better the crop in the following year.
- Plant strawberries in rectangular blocks rather than rows – that way they are easier to net.
- Newly planted strawberries are often overrun by ants, so keep an eye on your plants. A drenching of water usually helps.
- Cut down the foliage of perpetual ('everbearing') varieties in early autumn after fruiting has tailed off. Put the foliage in the bin.

VARIETIES OF PERPETUAL STRAWBERRY ('EVERBEARERS')

'Mara des Bois'
Dark-red, medium-sized berries with an alpine strawberry flavour. Crops well and resists mildew.

'Flamenco'
Heavy- yielding, new strawberry with large, firm, sweet fruit that keeps well on the plant. Resistant to verticillium wilt and powdery mildew.

'Finesse'
An even newer variety with heart-shaped berries on long trusses that stand proud of the lush, dense foliage. They ripen evenly and are also easy to pick. Crops from late June through to late August without producing many runners.

For further varieties of strawberry, see February, page 42.

2 Feed Blackcurrants, Pears and Plums with Nitrogen

(early March)

NITROGEN encourages leafy growth rather than fruit, so generally it is applied to fruit trees and bushes only if they are greedy feeders. Blackcurrants, pears and plums need a boost in early spring and the easiest way to apply nitrogen with certainty is to use hoof and horn or a general fertilizer. Apply 100g per square metre (3oz per square yard).

These nitrogen-hungry plants all like spring and summer rainfall. If the spring is dry, water them carefully in clement weather by gently tipping a bucket or two of water on the roots a couple of times a week. Always try to keep the water away from the trunk or woody stems – don't splash a hose on them. Do bear in mind that March nights are cool, so any watering should be over and done with by midday.

Did you know? The blackcurrant was first recorded in 1611, growing at Hatfield House in Hertfordshire. The gardener, John Tradescant the Elder, ordered twelve plants from Holland for the Earl of Salisbury's garden. The fruit was generally disliked by most people, however, and did not become popular until the late nineteenth century. The arrival in 1936 of the vitamin C drink Ribena led to mass cultivation, mainly in Herefordshire. Today 30 per cent of the UK's blackcurrants are grown in Herefordshire and most find their way into blackcurrant cordial.

SECRETS OF SUCCESS WITH BLACKCURRANTS

- Blackcurrants are shallow-rooted and struggle in droughts. They prefer cool conditions and rich, heavy soil that holds the moisture. They also enjoy full sun – although they will tolerate light shade.
- Plant 1.5m (5ft) apart. After planting always cut back all the shoots to about 2.5cm (1in) from the ground.
- Water blackcurrants during dry periods in the growing season.
- Feed with nitrogen in spring.
- Hand-weed and mulch around the plants to keep the weeds down and moisture in. Blackcurrants resent competition.
- Prune in late summer – see page 229.

VARIETIES OF BLACKCURRANT

'Ben Connan' ('Ben Sarek' x 'Ben Lomond')
High yields of exceptionally large fruits on a compact bush. Resistant to mildew and leaf-curling midge. Pick mid-July.

'Ben Lomond'
Flowers and fruits late, so will miss the frosts. A heavy crop of large, sweet berries on a compact bush. Mildew-resistant. Pick in late July.

'Ben Sarek'
Compact choice for the smaller garden. Frost- and mildew-resistant, so a good choice for colder sites. Heavy crops of large fruits ready for picking in mid-July.

'Wellington XXX'
This 1913 variety is a vigorous, spreading bush, producing a bumper crop of thick-skinned, sweet, juicy berries, even in hot summers. Pick in mid-July.

Organic Tip ✔

Blackcurrants flower early, so avoid planting in frost pockets – frosts can drastically reduce yields, even on some later-flowering modern cultivars. The Scottish-bred varieties with the prefix 'Ben' often flower late enough to miss the frosts.

3 Prune Redcurrants and Whitecurrants

(mid-March)

REDCURRANTS and whitecurrants are related to the blackcurrant (*Ribes nigrum*) but are derived from two different species, *R. rubrum* and *R. spicatum*. They fruit in a different way, on one-year-old wood and on spurs from older wood, so a permanent framework needs to be preserved. They do not need a procession of young wood like the blackcurrant, so shoots are not cut out at the base.

The objective of pruning redcurrants and whitecurrants is to create a goblet-shaped bush with between eight and ten main branches growing from a stem about 15cm (6in) off the ground. It is traditional to prune them as the sap rises, just as the buds begin to swell. Also, any buds that have been pecked and damaged by birds can be identified once the swelling starts.

Established bushes should have their leaders tipped by removing the top 5–7.5cm (2–3in) to stimulate new growth. All the laterals (side shoots) are then pruned back to one bud and any low branches are cut away at the source. This is the exactly the same method as used for gooseberries (see page 10).

Redcurrants and whitecurrants can suffer from die-back (a fungal disease). If the wood looks brown and dead when you cut into it, make a cut lower down where the wood is fresh and white. Remove it at source if you need to. If successive cuts reveal that die-back has entered the main stem, sadly the whole bush will need to be removed.

Did you know? The word 'currant' derives from the Greek city of Corinth and describes a small dried grape. Currants were even called 'corinths' in some early English books. The botanist John Parkinson, writing in 1629, is at pains to explain the difference between the dried currant and the fresh currant before extolling their virtues.

SECRETS OF SUCCESS WITH CURRANTS

- Redcurrant bushes thrive in open, sunny positions. However, they are tolerant of shade and they will fruit to a lesser degree on a north-facing wall – although later in the year.
- Plant bushes 1.5m (5ft) apart. They are more tolerant of drier soils than blackcurrants, so lighter dressings of manure (up to 5cm/2in) are best.

- Avoid frost pockets and exposed windy sites, although redcurrant and whitecurrant flowers seem to survive frost.
- Redcurrants crop very heavily, even on poor soil, and one mature bush usually produces plenty – up to 4.5kg (10lb) of fruit. Tie the branches to canes set around each plant to prevent the branches flopping to the ground under the weight of the fruit.
- Net the bushes tightly against bullfinches, blackbirds and thrushes before the fruit starts to ripen.

Organic Tip ✔

Red-, white- and blackcurrants are insect-pollinated. They can self-pollinate but crops are better when insects do the job. Currants (and gooseberries) flower early – beginning in March – at a time when not many flying insects are around. The main agents of pollination then are bumblebees and mining bees. These are quite large insects, so if you haven't lifted the side netting of your fruit cage to allow them access, make sure that the netting has a mesh large enough to allow them through. You will help our endangered bees as well as improving your crop.

VARIETIES

'Jonkheer van Tets' AGM
One of the earliest, bearing heavy crops of large red berries with an excellent flavour. Ripe in July.

'Red Lake'
A mid-season redcurrant producing large, heavy yields of long trusses of juicy berries that are easy to pick. Pick in late July.

'Redstart'
This late-season redcurrant produces heavy yields on an upright bush. It has excellent disease-resistance and its late-flowering habit avoids any frost damage. A slow-growing variety that makes a good choice for smaller gardens. Pick in August.

'Versailles Blanche' (syn. 'White Versailles')
Reliable yields of large, yellow-white berries. Good crops year after year. Pick in early July.

4 Plant an Apricot Tree

(late March)

SPRINGS are coming earlier and earlier, so gardeners should be taking advantage of the situation. This, combined with recent breeding (see Organic Tip, overleaf) is making it possible to grow apricots outside in the open rather than up against a warm wall. Commercial orchards planted 15 years ago are now cropping well and making a profit. Now that the ground is beginning to warm up it is the perfect moment to order one of these hardier apricots and to prepare the ground for planting.

The new varieties succeed in well-drained, open ground on good soil. A 5-year-old tree could produce 500 fruits in a sheltered position.

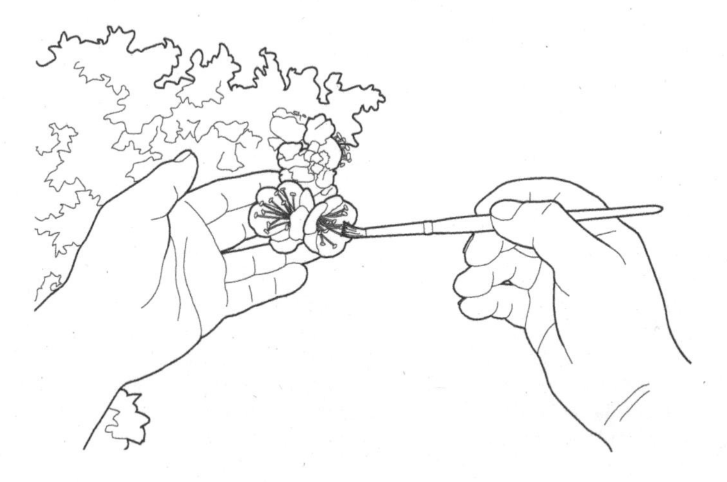

Apricots are self-fertile, so you will need only one tree, but you will have to discipline yourself to hand-pollinate the flowers, because generally apricot blossom arrives early, often before the bees are about to do it for you. It is traditional to use a rabbit's tail for hand-pollination, but a fluffy paintbrush is equally effective. Gently brush each flower when it is fully open (when the pollen is ripe). You won't see it, but pollen is transferred from flower to flower. Be methodical, moving along each branch, and dream of sun-warmed apricots while you do it. A home-grown apricot is utterly different from a bought one. Hand-pollinate on warm, sunny days.

Apricots don't suffer from as many diseases as peaches; bacterial canker and silver leaf are not serious threats. They can be pruned in late winter or early spring if you wish.

Organic Tip ✔

Apricots flower in February and growing them went into a decline in Britain largely due to colder winters. In recent years, however, North American- and Canadian-bred varieties (often with the suffix 'cot') have arrived. These trees can fruit in Britain without a warm wall and commercial production of apricots is the proof. Sainsbury's snapped up the first Kent-grown crop harvested in 2005. Grow one of these modern varieties.

VARIETIES

'Tomcot'
Produces a very heavy crop of large, crimson-flushed fruits which are ripe for picking from around the middle of July. Can be grafted on to Torinel rootstock, which is the latest rootstock for growing apricots. Approximate tree height when mature: 3m (10ft). (See also page 138.)

'Doucouer'
A French variety introduced in 2004, with a good flavour. Height at maturity: 3m (10ft).

'Goldcot'
Recommended for cooler, wetter climates. Hardy, vigorous and resistant to leaf spot, producing good crops of medium–large, golden-yellow, freestone fruit (the stone is not attached to the flesh – a great convenience in the kitchen) in August which will keep in the fridge for several weeks. Height at maturity: 3m (10ft).

Did you know? The apricot almost certainly originated from China, where it still grows wild in the mountains close to Peking. It spread along the Silk Road from east to west and Alexander the Great was said to have brought the fruit from Asia to Greece. From there it travelled to Italy, where Pliny wrote about in 1 BC. It was also grown and dried in Ancient Egypt. In Britain it became a gentleman's fruit in the late seventeenth century. 'Moorpark' (named after Lord Anson's garden near Watford in Hertfordshire, but often listed as 'Temple') was available by 1777 and continued to be the most widely grown variety for 200 years. Jane Austen even mentioned this fruit in *Mansfield Park*. A warm wall was required and Aynho, a village on the Northamptonshire/Oxfordshire border, boasted a particularly large number of trees on south-facing cottages. The original trees were brought there from Italy by the local squire, Sir Thomas Cartwright (1795–1850), and some still exist today.

SECRETS OF SUCCESS

- Prepare the soil well and incorporate lots of organic matter. Stake and secure with a tree tie as you plant, angling it at 45 degrees to avoid damaging the roots. See page 8.
- Water well in the first growing season.
- Summer warmth is essential for apricots, so find a warm, sheltered position.
- Be prepared to cover your tree when it's in flower if frost is forecast. A fleece blanket is ideal.
- Fruit forms on short spurs that are 2–3 years old, so give your tree at least 4–5 years to produce lots of fruiting wood.
- Thin clusters of apricots down to doubles.

5 Plant a Cherry Tree

(late March)

CHERRIES deserve a wider audience. New breeding and better rootstocks have arrived over the last 40 years or so and this is an excellent time to plant one or more of these recent varieties. Many are now grown on Gisela rootstock, which produces a 3–4m (8–10ft) tree that will fruit after only 3 or 4 years.

Plant cherries in a sunny site on average, well-drained soil. The shallow rootstocks used for all cherries mean that they do not survive in waterlogged ground: they pick up soil-borne root diseases like phytophthora. They also need permanent staking – again due to those shallow roots.

Canadian and British breeders have produced more cold-tolerant varieties that will crop in cooler temperatures, providing white blossom, fruit and autumn colour. They are mostly self-fertile too, so one tree will crop on its own. This makes them an ideal tree for a small garden.

There are two types of cherry. Sweet cherries (*Prunus avium*) need full sun and are generally eaten raw when ripe. Acid cherries (*P. cerasus*) are darker in colour and are normally cooked. They can be grown in shade against a wall. Both types will need a sheltered, warm site. This will attract the bees and, hopefully, protect that early-spring blossom from frost. Acid cherries crop all along the length of one-year-old wood. Sweet cherries crop at the base of one-year-old wood and on older wood.

Did you know? The Japanese revere cherry blossom and hold special ceremonies every spring during a national holiday calculated to coincide with the blossom. The cherry orchards of eighteenth-century England were also popular with tourists. Paddle-steamers would ferry visitors up the River Tamar from Plymouth to see the orchards on the slopes of the valley when the blossom was in its full glory. St Mellion, on the Cornish side of the Tamar, still holds an annual cherry feast every July. One famous cherry, 'Waterloo', was raised by Thomas Andrew Knight of Downton Castle in Herefordshire, the producer of the Downton Strawberry (see page 40). This fine dessert cherry fruited within weeks of the battle in 1815 and is still available – just.

VARIETIES

'Stella' AGM
An older, self-fertile variety with dark-red to black fruit. Early and prolific, it makes a good pollinator for other cherries. Pick in late July. Pollination Group D.

'Sunburst'
A new Canadian-bred, self-fertile dessert variety that produces almost black cherries. These are very large and full of flavour. Not as heavy-cropping as 'Stella'. Pick in July. Group D.

'Summer Sun' AGM
A very hardy variety raised in Norfolk and suited to cold areas. Produces a heavy crop of dark-red fruit with an excellent flavour on a naturally bushy tree. Pick in late summer. Self-fertile. Pollination Group D.

'Celeste'
Perhaps the best variety for a small garden because this compact, almost-dwarf, self-fertile cherry can be pot-grown. It was raised in Canada in 1990 and the dark-red fruits are very good to eat. Ripens early. Pollination Group B/C.

Organic Tip ✔

Cherries enjoy radiated heat and light coming up from the ground or off a wall. Place a light-reflective covering on the ground straight after flowering – use white stones or a man-made covering. The reflected light will help pollen-tube growth.

SECRETS OF SUCCESS

- Plant trees or bushes in a sunny site on average–rich, well-drained soil. If your soil is thin, improve it before planting by adding organic matter.
- Protect cherry flowers from frost by covering your trees on cool nights with a double layer of horticultural fleece supported on tall canes, keeping the fleece well above the blossom. This should allow pollinators access to the flowers during the day. Remove the fleece once the blossom has been pollinated and the petals drop.
- Give your tree a very light, sympathetic pruning after harvest – see page 208.
- Keep cherries well watered in the early stages of fruit development; otherwise all the young fruit may fall off in May. Cherries are self-thinning, so leave them to it.
- Feed in mid-spring with a top-dressing of general-purpose fertilizer.
- Few fruits attract as much attention from the birds as cherries do, so you must net. It's easier to net a cherry growing against a wall, if you have one.

6 Plant Raspberries and Other Cane Fruit

(late March)

Now is the best time to plant new raspberry canes, but do avoid those anonymous plastic bundles you can buy in garden centres; they may have been sitting there for months in airless, dark, damp conditions. Make sure that you buy certified stock from a good nursery and plant in clement conditions, avoiding frozen or waterlogged soil.

Dig in 10cm (4in) of organic matter 30cm (12in) either side of the intended row and erect strong supports at the end. Then make shallow holes and space each cane 35–45cm (14–18in) apart. Spread the roots out and cover with 7.5cm (3in) of soil. Rows need to be at least 1.5–1.8m (5–6ft) apart to allow access. Once planted, always cut each cane back to 22cm (9in) in height to avoid wind rock. On heavy soil, canes can be planted on a raised ridge to improve drainage.

Other cane fruit can also be planted now, including loganberries and tayberries. Both are raspberry × blackberry hybrids. The loganberry is a large, prickly plant that occupies 3.5–4.5m (12–15ft) of space and the long raspberry-like fruit ripens between August and September. It has a sharp flavour and is best cooked or jammed. The larger, sweeter and more aromatic tayberry crops first and the fruit is good for pies, jam or eating fresh.

The tayberry was developed at the Scottish Crops Research Institute, Invergowrie, Scotland, by Derek Jennings and David Mason in the 1970s. It has never become a commercial success because the berries are difficult both to pick by hand and to machine-harvest. If you have room, it's worth planting one. The 'Buckingham Tayberry' is a new thornless variety with smooth canes.

Did you know? The original cross between a blackberry and a raspberry was made accidentally by James Harvey Logan of Santa Cruz, California, in 1881–83. Logan was trying to produce a better blackberry and, while attempting to cross two varieties, he planted both next to a raspberry, thought to be 'Red Antwerp'. The two blackberries, 'Texas Early' and 'Aughinburgh', produced fifty seedlings between them and one, with long red fruit, was named the loganberry. Logan's original hybrid was introduced to Europe in 1897. It proved to be productive but its sharp flavour was not universally popular.

Organic Tip ✔

Always remember that raspberries grow best in slightly acid soil that is well drained and moisture-retentive. The best way to create ideal soil is through the liberal use of bulky organic manure, which is itself acidic and improves both drainage and moisture-retention. Even more organic matter than usual should be added to dry or limy soils.

SECRETS OF SUCCESS

- Choose an open site to attract pollinators: this ensures a heavier crop. Good light will also encourage much sturdier, stronger canes.
- Avoid windy sites. The fruit bruises easily.
- Try to run the rows north–south to minimize shading.
- Clear the site of perennial weeds before planting and improve the fertility of the planting trench with organic matter.
- Plant canes in clement conditions, avoiding frozen or waterlogged soil.
- Cut back newly planted canes to 22cm (9in) to avoid wind rock.
- For summer-fruiting varieties, add sturdy upright supports at the ends of rows and spread wire between them (autumn-fruiting varieties are self-supporting).
- Once established, chop out any canes that wander – especially those that invade the space between the rows.
- If new canes come up with mottled foliage, dig them up: it is almost certainly caused by a virus.
- For advice on cutting back autumn-fruiting and thinning summer-fruiting raspberries, see pages 38 and 227.
- Raspberries come from northern Europe and they prefer cooler summers – which is why they often do well in Scotland. Mulching helps to keep the soil cool and moist. Partially rotted grass clippings make a suitable mulch.
- Pick your fruit on a dry day.

VARIETIES OF SUMMER-FRUITING RASPBERRIES

'Tulameen'
Conveniently follows on from the strawberries, producing large, tasty fruit. Crops well.

'Glen Ample' AGM
This mid-season variety produces heavy crops of very large fruit on strong, spine-free, upright canes.

'Glen Rosa'
A new mid-season, spine-free variety, producing an abundance of bright, high-quality, medium-sized fruit. Very disease-resistant, so ideal for organic production.

'Malling Admiral'
The popular favourite, due to its excellent flavour and dark-red fruit. Quite a tall cane, although completely spine-free. Probably the best choice if you are restricted to one summer variety.

For varieties of autumn-fruiting raspberry, see February, page 39.

VEGETABLE

1 Sow Early Peas
(early March)

ALWAYS assess the condition of your soil before sowing early crops. If the soil sticks to your boots it is too wet to plant or sow, so turn your attention to verges and edges. Using a moon-shaped cutter, angle the tool slightly outwards rather than straight down. This exaggerates the edge and makes it look deeper. It also improves drainage and helps to prevent weed seeds congregating in the groove at the bottom. Neatening the edges and weeding all furrows will improve the appearance of your plot greatly and prevent the weeds from self-seeding.

When sowing your peas take heed of the old adage 'One for the mouse, one for the crow, one to rot and one to grow' and use lots of seeds. Make a 22cm (9in) wide shallow trench and zigzag the seeds across it from side to side. Cover the seeds with 2.5cm (1in) of soil and then cover the soil with wire netting to keep birds and mice away. Add twiggy supports straight after sowing – hazel is best – setting them along each side of the trench, about 20cm (8in) apart and at an angle, so that their tops meet over the trench and the pea shoots will weave through them.

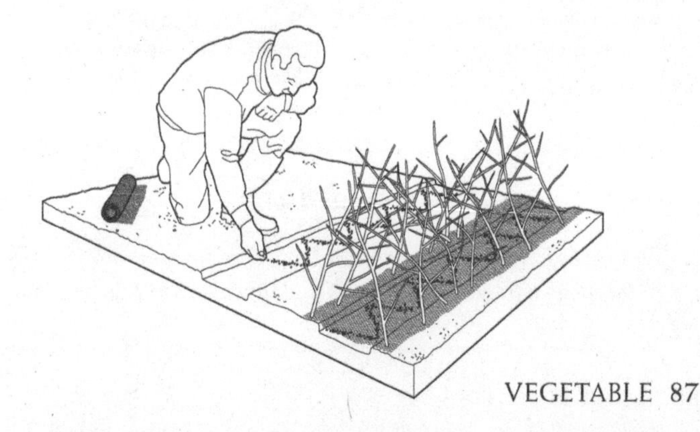

'Feltham First' is the best early pea variety and will reach only 40cm (16in). Mangetout varieties can also be sown now. Maincrops can be sown by mid-April. On average, peas take 100 days to mature and regular fortnightly sowings of taller, maincrop varieties can be made right up until late July.

EARLY VARIETIES

Paler, round peas with a floury flavour and grainy texture.

'Misty' AGM
A good cropper with blunt-ended, smaller pods each containing six peas.

'Early Onward' AGM
A heavy early cropper with pairs of pods.

'Feltham First'
For autumn or spring sowing, this round-seeded variety produces a substantial crop.

MANGETOUT VARIETIES

The pods are picked and eaten young – before the peas develop.

'Oregon Sugar Pod' AGM
Matures first and produces sweet, juicy pods for lightly steaming or stir-frying.

'Delikata' AGM
Taller and heavier-cropping than 'Oregon Sugar Pod', but gets stringy quickly.

MAINCROP VARIETIES

Sweetly succulent, bright-green peas that melt in the mouth.

'Jaguar' AGM
Ready after 100 days with short pods containing seven peas.

'Cavalier' AGM
British-bred, with pairs of long, straight pods containing nine peas. Good flavour.

'Hurst Green Shaft' AGM
Tried and tested heavy-cropper that produces long, easily picked pods. Rarely fails.

Organic Tip ✔

Peas (like all legumes) should never be given a nitrogen-rich feed. They fix their own nitrogen by forming an association with certain bacteria.

Did you know? Marrowfat peas seem as English as fish and chips, but they began life in Japan, hundreds of years ago, as the 'Maro' pea. Japanese growers thought the cooler climate here would make the peas fatter and the British-grown crop became known as Maro-fat.

SECRETS OF SUCCESS

- Peas like cool, moist conditions and often do best in cooler summers.
- They should be watered well at least once a week in dry weather as soon as they come into flower.
- Pick regularly to encourage more pods.
- Always sow a late crop in late July – they often do very well.

2 Sow Globe Artichokes

(early March)

THESE HIGHLY ornamental Mediterranean sun-lovers produce edible flower buds which can be harvested in early summer when little else is available to the vegetable gardener. Pick before the blue petals start to show, then simmer in water and douse them in butter for a quick and appetizing 'plot-to-plate' lunch. Alternatively, leave them alone so that they develop their cobalt-blue, bee-pleasing flowers.

Globe artichokes, commonly called cardoons, often succumb in cold winters. By early March you will be able to tell which of last year's crop have survived and which have died. Dig up any dead plants, clean away the debris and fill the gaps with offsets removed from your survivors. Use a sharp knife to slice the new outer shoots off below the ground. Each viable piece should have some root. Pot them up and give them 6 weeks in the warmest, lightest place you have. Bed them out once well rooted.

This is also a good time to think about sowing artichoke seeds. Sow two seeds into each small pot of compost; once germinated, select the best and bed out. If you have unheated glass you can do this in early March, or you can wait for 4 weeks and place the pots outside. Young plants can also be ordered now for May dispatch.

VARIETIES

'Concerto' F1
A new, vigorous variety with purple-washed, jade-green heads.

'Green Globe Improved'
Available as seed, this prolific variety produces less prickly heads.

'Imperial Star'
A highly selected seed strain from 'Green Globe'.

'Violetta di Chioggia'
Deep-purple-headed variety. Early, very ornamental and ready by June.

'Gros Vert de Laon'
A heritage French variety with the largest green heads and the best finely cut foliage.

SECRETS OF SUCCESS

- Globe artichokes demand a sheltered, sunny position and well-drained soil.
- These tall plants also need space to do well: thinning the stems to two or three per plant can make them easier to manage and less likely to topple over.
- Stake in windy gardens.

Organic Tip ✓

The best way to keep globe artichoke plants from one year to the next is to mulch them with straw in late autumn. Remove by February because these handsome, statuesque plants start into growth early. Avoid planting in frost pockets, as the new foliage is vulnerable. Remove the faded stems at the base in summer to encourage a late flush.

Did you know? The Greeks and the Romans both ate globe artichokes. The names *kardos* (Greek) and *carduus* (Latin) both translate as 'thistle'. The Romans believed this plant had aphrodisiac qualities and they imported them from Cordoba in Spain in large numbers. Henry VIII grew them at New Hall in Essex in 1530 – possibly for the same reason – but in his day women and people of low birth were not allowed to eat them. They were literally forbidden fruit.

3 Sow Three Varieties of Lettuce

(mid-March)

SELECT THREE different types of lettuce to extend the picking season. Choose a loose-leaf 'pick and come again' variety like 'Salad Bowl Mixed': this soft mixture of red and green oak leaves will be ready to pick after 8 weeks. Sow a small, hearting lettuce like 'Little Gem' or its red-leaved equivalent, 'Dazzle': these will be ready to cut 2–3 weeks later. Finally, sow a slower-maturing Cos variety like 'Lobjoits Green', which often takes 10–12 weeks to fill out.

Always use fresh packets of seeds with the correct date because lettuce has a short period of viability – 3 years at most. Write out the labels before you start and fill the trays with compost to within 1cm (½in) of the rim. Water well with mains water before sowing. Ideally, the water should stand in a fine-rose can for at least half a day to warm up and release some of its chlorine. Using tapwater (i.e., not water from a water butt) prevents damping off – a fungal disease.

Sprinkle the pale seeds very thinly on the compost and cover lightly with a fine layer of compost. Place in a cold frame, or in an unheated greenhouse, or on a cool windowsill. Ideally, seeds will germinate within 6–14 days in reasonable temperatures.

Prick out when two proper leaves show and then plant outside once large enough. Repeat the process every 4 weeks until late July to ensure a long supply of salad leaves.

VARIETIES

'Little Gem' AGM
The best early small lettuce. It hearts up well.

'Dazzle'
Similar in shape to 'Little Gem' but with burgundy leaves.

'Salad Bowl Mixed' AGM
An early cropper, this decorative green-and-red oak-leaved lettuce is a loose-leaf variety.

'Lobjoit's Green Cos' AGM
Large, crisp, green-leaved Cos. Suitable for spring and autumn sowing.

'Nymans'
A medium-sized, shiny-leaved, red Cos lettuce. Slow to bolt (run to seed).

'Romaine'
Survives hot summers and produces crisp, dark-green, solid hearts. Traditionally used in Caesar salads.

Organic Tip ✔

Lettuces attract slugs, but if you grow African marigolds close to them or among them the marigolds act as slug magnets. Collect them at dusk and exterminate!

Did you know? The milky sap that exudes from a cut lettuce stem contains a narcotic similar to laudanum. Hippocrates, who was born on the Greek island of Cos in 460 BC, recognized that lettuce had a soporific effect. The long-leaved Cos lettuce almost certainly originated on this island – hence the name. Beatrix Potter also noted lettuce's soporific effects in *The Tale of the Flopsy Bunnies*!

SECRETS OF SUCCESS

- Lettuce is a cool-season crop and most varieties struggle to germinate and grow in hot conditions.
- Sow on cooler days, or in the evenings if the weather's hot.
- Grow a selection. Lettuces mature at different rates and some (such as Cos) varieties are much less likely to bolt. Once planted out, always water in dry weather to prevent bolting.

4 Sow Tender Vegetables under Cover

(mid-March)

TENDER VEGETABLES like aubergines, chillies, peppers, squashes and tomatoes to grow outdoors can be sown under cover now. Put them in a greenhouse or on a sunny windowsill and keep the pots warm. Ideally, aubergines and peppers need temperatures above 15°C (59°F) in order to germinate within 10 days.

Large cucurbit seeds (like cucumber, courgette and squash) should be sown vertically (with the sharp end down – the root emerges from the pointy end), as they can rot easily if laid flat in the compost.

These easily handled seeds can be sown straight into pots. Smaller seeds (like tomatoes and peppers) can be sown in pots or trays and then pricked out as soon as two true leaves have appeared. Use multi-purpose compost or John Innes No. 2. Loosen with a small dibber and handle by the leaves, not the stem. Firm them in gently and water well. Thereafter water sparingly to encourage deep root systems. Use warm cans of tapwater (not from a water butt) to prevent fungal diseases such as damping off. Each seedling should take 5–8 weeks to develop into a proper plant.

Keep seed trays under cover (or somewhere very sheltered) until early June, and cover with thick horticultural fleece if a cool night is forecast. Remove it during the day.

As soon as plants come into flower, start to water on a high-potash tomato feed every 2 weeks to encourage flowers and fruit. Keep feeding your plants until the beginning of September to keep them productive.

SECRETS OF SUCCESS

- Frost-tender plants need raising in the greenhouse, or buy them from garden centres. Harden them off carefully before putting outside.
- Plant outside once the fear of frost has passed. June is always better than May because the slightest frost can blacken and kill young courgettes, cucumbers and squashes. Cold nights will also check the growth of tomatoes, peppers and aubergines.
- Water all these plants regularly in the early evening or early morning.
- Feed aubergines, peppers and tomatoes with a high-potash tomato feed every 2 weeks. Cucurbits do best on pelleted chicken manure.
- Squashes must be stored for 6 weeks before you eat them. This allows the starch to turn to sugar.

VARIETIES

Aubergine
'Bonica' AGM
The easiest variety, producing large, dark fruits. Needs a bumblebee for pollination.

Chilli Pepper
'Hungarian Hot Wax' AGM
Conical yellow fruits that ripen to red. Good to eat raw and also for cooking, although flavour intensifies.

Courgette
'Romanesco'
Nutty Italian courgette with ridged green fruits. 'El Greco' AGM (smooth, non-prickly, dark green), 'Soleil' AGM (slender yellow) and 'Venus' AGM (compact green) are also excellent.

Squash
'Sunshine' (All-American Winner)
A small (up to 2kg/4lb), orange, pumpkin-shaped winter squash with a chestnut flavour – similar to 'Potimarron' (up to 4kg/8lb) and 'Uchiki Kuri' (2.5kg/5lb). All three are much easier to grow than butternuts.

Sweet Pepper
'Bell Boy' AGM and 'Gourmet' AGM
Both block-ended, traditionally shaped peppers that turn from green to orange.

Tomato
'Gardener's Delight' AGM and 'Sungold' AGM
Both form large trusses holding lots of cherry tomatoes. 'Gardener's Delight' is red and 'Sungold' a sweet orange-yellow.

Tomato
'Tigerella'
Striped red-and-green tomato with a top flavour.

Tomato
'Beefsteak'
Large, fleshy tomato, good to eat raw or for cooking. Heavy yields in warm summers.

For varieties of courgettes, cucumber and squash, see page 131.

Organic Tip ✔

Water the pots of compost well before sowing these seeds and then water very sparingly, taking care not to soak the pot again, until the seeds have germinated. Once the seedlings appear, water in the morning (again sparingly) and fleece overnight to prevent night-time chill.

Did you know? Peppers have been grown and eaten for thousands of years and they are depicted on the Tello obelisk, a granite shaft carved by Peruvian Indians between AD 800 and 1000.

5 Sow Carrots
(late March)

HOME-GROWN carrots eaten straight from the garden are sweet and nutty. They are poles apart from shop-bought ones so are worth growing yourself. Carrots germinate quickly once spring arrives and temperatures reach 12°C (54°F) and above. However, they fail if it's colder, so sow in March only if the weather is on your side. Your first crops should be harvestable from mid-June onwards, but some hardy varieties can be dug during winter. You can succession-sow every 14 days up to late August – although late sowings don't always succeed.

Cultivate the soil deeply before sowing, removing any stones with a rake. You can either sow rows or use the handle of a rake to make 15cm (6in) wide, 2.5cm (1in) deep trenches or 'drills'. Aim to sow thinly so that each carrot has its own space to develop and thinning will not be necessary. If you have to thin, do it on a damp day. Less disturbance will lead to fewer problems with carrot root fly because the aroma of exposed carrots attracts them. Harvest carefully and always cover up any exposed roots afterwards.

Varieties differ in root shape and colour. If you have stony soil, grow a stumpier carrot like 'Chantenay Red Cored' – one of the best earlier varieties. Cloching in early spring helps to warm the soil.

VARIETIES

'Early Nantes 2'
Long, tapered roots early in the season. A staple variety. 'Valor' AGM is a Nantes-type hybrid.

'Amsterdam Forcing 3' AGM
Smooth, blunt-ended carrots. Strong, with short foliage. The earliest variety of all.

'Chantenay Red Cored'
Red, coreless carrot with stumpy roots. Does well on stony soil.

'Purple Haze'
Smooth-skinned, slender, purple-skinned variety with a yellow middle. Does well in hot summers.

'Kingston' AGM
A late hybrid carrot for winter storage. Good colour and flavour.

'Eskimo' AGM
The best at overwintering due to its shallow crowns which stay under the ground.

SECRETS OF SUCCESS

- Carrots are umbellifers and this family (which includes parsley and parsnip) germinate only in warm conditions, so always wait for the weather.
- Make a drill and water it well. Sprinkle the seeds thinly and just cover lightly with soil. Protect with netting.
- Leave 30cm (12in) between each row, or make 15cm (6in) wide drills that don't need any thinning.
- Don't sow on newly manured ground – the carrots will fork.
- Rotate your crop to prevent a build-up of pests.
- Clear all carrots before the beginning of January as they can harbour slugs and other pests.

Organic Tip ✔

Carrot root flies hover close to the ground and you can protect your crop with 30cm (12in) high mesh screens – although you will have to water your carrots more, as the mesh keeps off rain. The classic sign of carrot infestation is crimson-red leaves, but don't pull them up if you suspect damage. Leave any affected roots in situ. *If you are plagued by carrot root fly, later sowings, from early June onwards, may be better.*

Did you know? Carrots grow on poor, sandy soil in temperate zones all over Europe and Asia. Root colour varies in wild populations from white, purple and yellow through to orange. Dutch plant-breeders chose to cultivate orange carrots from the sixteenth century onwards to honour the House of Orange, and all orange varieties are high in carotene. Purple- and red-rooted carrots grow naturally in the Hindu Kush region of central Asia (possibly the home of the carrot) and they are more drought-tolerant and rich in anthocyanin.

6 Plant Early Potatoes
(late March)

PLANT SOME of your first early potatoes in the third week of March for a late June crop. Be prepared to protect them with thick horticultural fleece at night, however, as the slightest frost will kill off the foliage and ruin the crop.

Earth them up as a further protection. Plant the rest of your earlies by mid-April and by the time they pop through the ground the frosts should be over for the year.

Varieties range in flavour and texture from the floury to the waxy, and there are many to choose from. Some (like 'Rocket' and 'Swift') produce a crop in 10–12 weeks, but generally these fast-maturing varieties produce large, tasteless potatoes if left in the ground for longer. Waxy varieties boil well and have a distinctive flavour. Floury varieties can disintegrate when boiled and may be better steamed. Bear in mind that all these potatoes need eating fast. They will only store for a matter of weeks.

The big advantage of early potatoes is that they are easy to grow and full of flavour. They are also generally out of the ground before mid-August when potato blight usually strikes, egged on by humid conditions. Once the potatoes have been dug up, water the ground well and add a slow-release general fertilizer (such as blood, fish and bone), then plant another, different, crop such as leeks, cabbages or dwarf French beans.

VARIETIES

For varieties, see January, page 25, August, page 242, and September, page 281.

Did you know? Potatoes are natives of the high Andes and were found by the Conquistadors in 1537, then introduced into Europe in 1570. Consequently they weren't mentioned in the Bible and so were believed to be the devil's food. Catholic Ireland got round the problem by sprinkling them with holy water and planting them on God's Friday – the old name for Good Friday – a religious superstition that still prevails.

SECRETS OF SUCCESS

- Always buy early potatoes early enough to chit them. Lay them out on egg or seed trays in a light, frost-free shed so that they produce strong, short shoots. This chitting process speeds up the crop. Maincrops do not need chitting.
- Try to prepare the soil well and always plant the tubers in damp soil that has begun to warm up.
- Rotate potatoes on a 3- or 4-year system to prevent eel-worm.
- The first sign of blight (*Phytophthora infestans*) is flagging green foliage followed by spotting. If it strikes, cut back the foliage to minimize the spread and destroy it. Don't add it to the compost heap. Some potatoes have more blight-resistant tubers. For blight-resistant varieties, see page 242.

Organic Tip ✔

Chop off comfrey leaves (preferably 'Bocking 14') and place in the bottom of the potato trench before planting. As the leaves decompose they will boost the nitrogen in the soil and also add potash and potassium.

APRIL

Although April is meant to be the first of Robert Herrick's 'four sweet months', it can often be fickle. Sometimes it lives up to T. S. Eliot's description of the 'cruellest month' of all. This is the prime month to sow hardy crops such as beetroot, spinach, wrinkle-seeded peas and broad beans out of doors. You may still be playing the waiting game in colder districts, but once you see a growth spurt, ride the wave as enthusiastically as a surfer on a roaring swell.

Carry on getting the soil ready by raking it down into a fine tilth whenever the weather allows. Once the texture is fine enough, leave the soil to stand and settle for at least two or three days – particularly where finer seeds are to be sown. If April is still chilly, cloche or cover the areas where you want to sow because warm soil is essential for fast germination. If April is warm, capitalize on every moment and sow and plant all your hardy crops in earnest.

In the fruit garden the earlier flowering tree fruits (such as plums and pears) begin to show blossom by April, but late frosts can ruin the blossom and result in a poor crop, or no crop at all.

You're in the lap of the gods, because we can't predict the weather, but we can limit its effects because there are differences in temperature within every plot. Cold air always falls to the lowest point and this is why so many orchards are on south- or west-facing slopes: the trees escape damaging frosts and benefit from better drainage. Never plant fruit in a dip, or at the foot of the slope, where cold air is likely to collect. Light levels also have to be good, because warm sunshine makes the nectar flow and lures in bees and other pollinators. Warmth also promotes the growth of pollen tubes and this is particularly important with pears. They do not do well in cool positions and this is why so many pears are trained on warm walls.

FRUIT

1 Fleece Strawberries
(early April)

If April is warm and sunny, early varieties of strawberries will begin to flower and this puts them in the danger zone. If the flowers get frosted they turn into black-eyed Susans (so they are very obvious) and they never bear fruit. Your crop is lost. Frosts are commonplace in April – they can even strike as late as the second half of May – so strawberries always need protection. Pay special heed to the weather forecast and prepare to roll out the fleece as needed. Horticultural fleece, one of the very best modern inventions, is inexpensive, light, and it does its job well.

You may also like to continue the war on slugs by watering on slug nematodes (tiny parasitic worms) around your strawberries. If so, wait until late April when the soil and air temperatures are warmer. Research has shown that a warm, damp day is best and the optimum time is 4pm, when the April sun is cooling off. You won't see lots of dead slugs, as everything happens underground, but the effects last for 6 weeks or so and, when the slugs are gone from the plot, the nematodes also die.

Did you know? Looks are not everything when it comes to strawberries. The best-looking strawberries are produced by the Indian strawberry, *Duchesnea indica*. Named for Antoine Duchesne, a Frenchman who somehow managed to write a monograph on strawberries in the middle of the French Revolution (see also page 40), its fruit are perfectly shaped, perfectly red and utterly tasteless.

Organic Tip ✔

Slugs love strawberries, but slug bait is not eco-friendly. It can be very toxic to pets – especially dogs. It doesn't discriminate, but kills every slug, including garden-worthy ones. The larger black-and-orange slugs (usually forms of Arion ater*) prefer debris (or detritus) to young growth; they also clean up rotting leaves and fallen petals and do much good in the decomposition process. Killing these is not good garden sense. It tends to be smaller slugs that do most damage, not the large wrinklies.*

SECRETS OF SUCCESS

- For advice on planting and caring for strawberries, see also February, page 40; March, page 69; May, page 134; June, page 174; and August, page 230.

VARIETIES

For varieties of strawberry, see February, page 42, and March, page 72.

2 Avoid Spraying
(early April)

APRIL IS a key month for the birds in our gardens because most begin to nest now. Chaffinches, blue tits, coal tits, wrens, robins and blackbirds will have their broods to feed this month. All baby birds (even those of seed-eating birds) need a diet of invertebrates until they fledge. They are in the nest for roughly 21 days, during which time an average brood of seven blue tits will consume 10,000 insects and grubs – and there may be several nests in and near your garden. Gooseberries, fruit trees, currants, raspberries and strawberries will be frisked daily. When a bird finds a good source of food (young sawfly grubs on a gooseberry, for instance), she or he will ferry back and forth between there and the nest, cleaning up the whole colony.

This is far more effective pest control than spraying because you inevitably miss some. In time, insects also develop chemical resistance, so the spray becomes next to useless. Worse still, the balance of your garden is completely ruined. It relies upon a complex set of relationships between predator and pest – the ladybird needs the aphid, for instance.

Sprays exterminate both pest and predator – and this also happens if you use organic or 'green' soft soap, mustard and garlic sprays. However, the pest can bounce back quickly – in a few days – because most have short life cycles. Aphids, for example, can produce forty generations in a growing season, given the correct conditions. Predators' life cycles are much longer and most produce only one or two generations per year, so they take far longer to recover. With no predators, the pests get the upper hand and you get more problems, not fewer.

Relying on Mother Nature is the best option for the home gardener. This is what generations of gardeners did in the past – and it works!

Did you know? Red-breasted robins are very aggressive birds and will fight to the death over their territories. This is why they sing so persistently – to deter any incomers. Their most prized possession is the spindle tree (*Euonymus europaeus*) because the bright pink-and-orange fruits are the most nutritious of any hedgerow plant.

CREATING A GOOD ECO-SYSTEM

- Building up a good eco-system relies on how attractive your garden is to insect life, both above and below the ground.
- Make it a chemical-free zone.
- Plant diversely, using trees, shrubs, perennials, evergreens, grasses, ferns, annuals and bulbs to attract a wide range of insects.
- Have undisturbed areas – hedge bottoms, wilder banks and some long grass, for instance – where insects and small animals can hibernate.
- Plant different areas of the garden according to conditions – shade-lovers in a shady spot, moisture-lovers in damp areas, etc.
- The boundaries between differently planted areas are particularly good for insect diversity. Try to add new areas.
- Provide blossom from early in the year until late. Start with snowdrops, crocus and hellebores and end with winter-flowering shrubs like *Lonicera purpusii*.

Organic Tip ✔

About half of all aphid species target a certain plant. The large lupin aphid, for example, rarely goes on to other plants. If you see a colony in your garden and it really troubles you, don't use a spray – green or chemical – and don't try to spray them off with water. Instead, rub them through your fingers. Wear gloves if you are squeamish. Aphids have a fragile feeding tube called a stylet and once this is broken, they die. It is their Achilles heel.

3 Blossom and Bees

(*mid-April*)

THE IMPORTANCE of pollination cannot be stressed enough. Without it, you won't get any fruit.

Pollination simply means the transfer of pollen from the stamens, or male parts, of a flower to the pistil, or female part, of a flower. The pollen cannot, though, arrive on any part of the pistil; it must land on a specialized surface designed to receive pollen – the stigma. Once it has arrived, the pollen grows tiny, microscopic pollen tubes into the ovary and fertilizes the ovules – but only if the pollen and ovules are compatible. Compatibility is under genetic control; it works on the same principles as our own immune system.

Some fruit trees, such as nectarines, can be fertilized by their own pollen; they are self-compatible. Some, such as most sweet cherries, are self-incompatible. They must be cross-fertilized by pollen from another variety of the same fruit – it must be another variety and not just another plant of the same variety. This is because all the plants of a single variety are clonal – that is, they have been propagated vegetatively and are genetically identical. So, pollen that lands on the stigma of another plant of the same variety will be rejected because the stigma recognizes the pollen as itself. Other fruits, such as some apple varieties, show partial incompatibility: they can pollinate themselves at a pinch, but fruit is much better if they are cross-pollinated. These are best treated as fully self-incompatible because we want as much fruit as possible.

We can manage pollination in the fruit garden using a few simple rules.

The first is to know which fruits are self-fertile and which are not. You can have single specimens of self-fertile fruits but you may need to plant two varieties of self-incompatible fruits. For example, all apple varieties are divided into seven pollination groups according to when they flower, Group A being very early and Group G very late. All the varieties in one group flower at about the same time, ensuring that all-important cross-pollination. Flowering also overlaps partly with varieties in some adjacent groups, so a variety in Group C will very probably be pollinated by one in Group B or Group D. Even if varieties flower together, some pairings are not compatible, though. 'Cox's Orange Pippin' will not cross with 'Holstein' or 'Kidd's Orange Red', for example. Some very good apples, like 'Bramley Seedling', are triploid (they have three sets of chromosomes) and produce hardly any viable pollen, so they make ineffective pollen partners. If you wish to grow a 'Bramley', you will need to plant a trio, using two other varieties in order to cross all three. Other bountiful varieties are biennial – they produce a bumper crop every second year – which also means that they flower poorly every other year and they are ineffective pollinators in their fallow, poor years.

Apples, pears, some plums, blueberries and most cherries need cross-pollination. Cherries, in particular, have a complicated incompatibility system. However, new varieties are self-fertile, which helps a lot.

Besides the plants and their flowers, we also need to think about the agents of pollination. The vast majority of pollination in the fruit garden is carried out by insects and particularly by bees. The list of bee-pollinated fruit is almost endless: apples, pears, plums, gages, damsons, cherries, quince, peaches, apricots, strawberries, raspberries, blueberries, blackcurrants and many more. The exceptions are nuts such as hazelnuts and walnuts, which are wind-pollinated. They produce vast quantities of pollen and rely on luck and the four winds to transfer it to the stigma.

Insect-pollinated plants have learned to improve the chances of pollination by providing the pollinator with rewards of energy-rich nectar and protein-rich pollen. Pollen and nectar are positioned in the flower to maximize the chances of transfer when the pollinator visits. The pollinator moves from flower to flower, picking up pollen from stamens and depositing it on stigmas as it goes.

Bees are critically important to the fruit-grower. There are about 250 species in Britain and most of them visit flowers, pollinating as they go. They travel some distance when they forage; honeybees can travel 6km (4 miles) or more away from the hive, but journeys by other bees are much shorter. Solitary bees, such as mining bees and potter bees, may travel no more than 200m (220yd). Bumblebees can travel a little further – perhaps 1km (just over half a mile). So if you have space for only one variety of a self-incompatible fruit, you may be able to rely on pollen from trees in the neighbourhood, but some fruits, such as cherries, are not widely grown, so the chances of bees bringing pollen to your plant are much lower.

Did you know? We look on the honeybee as a model of industry, but bumblebees are far more efficient pollinators. They fly in cooler temperatures, so early-flowering fruit often relies on them rather than on honeybees. When visiting a flower, bumblebees frequently vibrate their wings violently to shake pollen free. This is called 'buzz pollination' because of the sound the bee makes.

Organic Tip ✔

Encourage wild bees into your garden. Springtime blossom is excellent for this, but to get them really visiting you need to provide flowers from early spring, when the pregnant queens emerge from hibernation, until the new young queens go into hibernation in autumn. Nest sites include holes in walls, mossy vegetation in hedge bases and even old mouse nests.

SECRETS OF SUCCESS

- Late-pollination groups are often recommended for frost-prone gardens; however, most popular varieties of apple tend to be in Pollination Group C.
- Use a reputable fruit-grower who knows a lot about varieties – including pollination groups. Size and vigour are restricted by rootstocks (see page 207), but variety also plays a part. Triploid apple trees (like 'Bramley Seedling') make large trees. Pears are fussier about their pollination partners than apples.
- In very built-up areas where there is a lot of top fruit in surrounding gardens, there is usually plenty of pollen. You may get away with planting just one fruit tree – but it's a gamble.

4 Grow Fruit in Containers

(mid-April)

THIS IS an excellent time to buy container-grown fruit trees and bushes. A lot of modern breeding has gone into producing dwarf varieties suitable for small gardens and, if your plot is tiny, these can be extremely worthwhile. The Californian plant-breeder Floyd Zaiger has bred a series of dwarf fruit trees that produce full-size fruit. These rounded trees can be grown in the ground and after 7 years will reach 1.5m (5ft) in height. Grown in pots, they will stay even smaller. Their diminutive size makes them ideal patio plants.

However, if you go down the container route you have to be extra vigilant about plant care. Growing in a container exposes plants to drier, hotter conditions in summer and colder conditions in winter. It is much more stressful for the plant than growing in the ground. Nutrients soon become scarce, so feeding and watering are important.

Most dwarf fruit comes ready containerized. If you want to re-pot it, use wood or terracotta: both are far kinder to the roots in summer and winter. Black plastic pots should not be used: the dark plastic absorbs the heat too efficiently and does not protect against frost. Aim for a pot 38–55cm (15–22in) in diameter.

The big difference between the care of container and open-ground fruit trees is that container fruit must be root-pruned occasionally to maintain fruit production. This is done in late autumn.

These dwarf varieties give every gardener the chance to raise some of their own peaches and nectarines. Pot-grown fruits are easier to protect *in situ* against frost, birds and high winds. Alternatively, you can move them into a cold conservatory or slightly warm greenhouse. Dwarf cherries and smaller fig varieties can also be grown in a pot. Figs should be taken inside in early winter.

South-facing walls are always cited as the best against which to place container-grown fruit, but in hot summers this can frazzle plants. I would advocate a west-facing or south-west-facing wall, but avoid north-facing walls: these are too sunless. Also avoid east-facing walls as they get the sun early and thaw out frosted flowers and buds far too quickly.

Did you know? Potted fruit is nothing new. Grapes were often grown in pots (using methods described by the garden writer John Claudius Loudon in 1835) and when the trusses were ripe, the whole plant was taken into the dining room, pot and all, so that guests could pick their own ripe fruit.

VARIETIES

Dwarf Cherry
'Garden Bing'
Large, dark-red clusters of cherries that are ready to pick in late July. The snow-white blossom appears in April. (Do not confuse with the full-size 'American Bing' varieties.) Self-fertile.

Fig
'Rivers Brown Early'
Large, pear-shaped brown fruit with a purple tinge. A reliable, early, abundant variety rescued from the famous Rivers Nursery near Sawbridgeworth in Hertfordshire when it was demolished for development. Self-fertile.

Nectarine
'Rubis'
Large, lance-shaped and colourful yellow fruit flushed with red. The pink blossom appears in March so, although self-fertile, you will need to hand-pollinate (see page 77).

Peach
'Bonanza'
Genetic dwarf peach, bred in California. It produces a 'mop head' of full-sized leaves, a mass of pink blossom and full-sized fruits. Self-fertile.

SECRETS OF SUCCESS

- Choose a rugged, weatherproof container and make sure that it has drainage holes. Add a layer of crocks at the bottom to assist drainage further and stand your pot on feet – this allows surplus water to escape.
- Use loam-based John Innes No. 3 for your compost. This retains moisture more efficiently than peat-based composts and delivers nutrients more slowly. Firm down well and always leave at least 5cm (2in) at the top to allow for mulching and watering. Add 2.5cm (1in) of well-rotted compost or manure and a little general fertilizer.
- Water regularly from March onwards. Start with a little at first and then increase the amount. Allow a little to run out of the bottom to stop the build-up of harmful salts.
- Feed with a high-potash (tomato) fertilizer fortnightly from April until late August to encourage flower and fruit.
- Also use a foliar seaweed feed every 2 weeks to strengthen and toughen the foliage. This prevents pests and diseases.
- Figs should be taken in during December.
- At leaf fall, remove all remaining figs larger than small peas. Leaving them on will reduce next season's crop and may cause die-back.
- Take out growing tips of all new-season shoots on peaches, nectarines, cherries and figs around the middle to end of June, a little earlier than you would with open-ground trees.

Organic Tip ✓

Pick your fruit when it's truly ripe. With peaches and apricots a gentle twist is usually all that's needed to detach the fruit if it is ready. With figs, the fruits droop downwards in late summer or early September and begin to change colour. Look for the drop of nectar at the base of each fruit – it's called the 'tear in the eye'. If you can see it, the fig is perfect for picking.

5 Beat the Spring Drought

(*late April*)

BRITISH weather is unpredictable and we can't do anything about varying or erratic temperatures, or sunlight hours. We just have to accept what we get stoically. However, we can conserve moisture by mulching and spring is often a good time to do this, as we can get a period of drought. Mulching will also keep weed growth down. Late April is ideal, because if a mulch is applied too early it tends to get washed away.

Mulching needs to be well timed: it should follow a period of heavy rain and the soil must be warm. In dry conditions you may need to soak the ground before applying a mulch. Those on lighter soil in the drier eastern half of England will benefit most; those in wetter areas may not need to bother. If you're on alluvial soil close to a river, the ground may already be very moist due to a high water table.

Mulches decompose and as they do so they use up nitrogen in the soil, which may cause plant stress. If well-rotted manure and garden compost are being used the decomposition process should be more or less complete, so there will be plenty of nutrients on offer from the organic matter. However, some mulches (like ornamental bark) deliver very little in the way of nutrition. If using bark or gravel, always add a nitrogen-rich feed before spreading the mulch.

Did you know? Warmth encourages the development of the pollen tube, so warm Mays are what fruit-growers pray for.

Organic Tip ✔

Pears are the most demanding fruit trees when it comes to moisture and they often do well in low-lying river valleys where the soil is rich. They also like warmth. Always water pears from the moment the flower buds burst until 6 weeks after blossoming to improve the yield. Gently tip one bucket of water on the roots of each pear tree every day in dry spells.

SECRETS OF SUCCESS

- Don't mound mulch up right against the trunks of trees and shrubs: it will kill them. Swirl a 5cm (2in) layer around them instead.
- Weed and aerate the soil before applying mulch. If it's a mulch that offers little or nothing (bark, gravel, sawdust, etc.), add a nitrogen-rich fertilizer first.
- The downside of mulching is that it can harbour slugs and snails – so mulching near slug-prone plants is not a good thing.
- Local authorities are producing green waste material from garden refuse collections. This should be worth trying for many, unless you are an organic gardener: it cannot be guaranteed chemical-free.

6 Remove Protective Coverings from Peaches and Nectarines

(late April)

MOST OLDER and heritage varieties of peach, nectarine and almond are prone to a fungal disease called peach leaf curl (*Taphrina deformans*). In early spring the leaves produce red or white blisters and then the whole leaf curls up before dropping to the ground. This devastating disease effectively defoliates the tree and starves it of food. Although a second crop of leaves follows (and they often stay healthy), the damage has already been done and the tree will fail to produce much fruit.

The fungus overwinters on the shoot tips of the plant, especially in the scales surrounding young buds. The fungus is transferred from the bud scales to the young leaves in spring by rain, where it causes the characteristic deformed leaves. The fungus grows, sporulates, and the spores are returned to the shoot surfaces to begin the cycle again. Most experienced fruit-growers opt for prevention. If you can keep the rain off the peach tree in winter the spores are not transferred to the young leaves. Cover the whole tree with glass or polythene over winter to keep the buds dry. Leave the sides open to encourage air flow. The covering also protects the early blossom from frost. Also, gather up and destroy infected leaves early so that the fungus cannot sporulate. See also page 306.

Now is the time to remove your cover and allow the tree to form healthy leaves. Older peach varieties are much more prone to peach leaf curl than newer varieties, and nectarines are more difficult to grow than peaches.

Did you know? Peach leaf curl is caused by just one of a specialized group of fungi. They all disrupt the plants' hormone systems and cause distorted growth. Pocket plums – distorted fruit with a hollow, pocket-like side – are caused by *Taphrina pruni*. Cherries and pears suffer attacks from other specialized *Taphrina* species.

Organic Tip ✔

Do not be tempted to use the traditional fix for peach leaf curl – Bordeaux mixture – even though it is still approved for organic use (though no longer recommended). It is very high in copper, which is toxic to many beneficial creatures, especially frogs and toads.

SECRETS OF SUCCESS

- Wall-trained trees generally crop much more reliably for most and it is worth buying a ready-trained, fan-shaped tree (see page 232).
- Always protect the blossom from frost. Hand-pollinate, because there are few (if any) pollinators about.
- Water regularly once the fruit begins to swell.
- Thin the fruit twice during the growing season – at hazelnut size in about May and then at walnut size in June – so that they are well spaced.
- Prune carefully (see page 208), taking some of the older branches out. Create an airy middle – this will lessen the chance of fungal diseases because the wood will dry out better.
- If leaves do curl, pick them up and bin them. Good husbandry is the best prevention.

VARIETIES OF PEACH RESISTANT TO PEACH LEAF CURL

'Avalon Pride'
A chance seedling found growing wild in woods near Seattle in 1981, this outstanding, breakthrough variety is strongly resistant to leaf curl disease. The large fruit is ready in August. If grown on 'St Julian' rootstock, it is less susceptible to frost damage.

'Red Wing'
This forms a large tree and yields well, producing almost completely dark-red fruit with a superb flavour. Late-flowering, so good for colder sites. Has some leaf curl resistance.

TRADITIONAL VARIETIES OF PEACH

'Peregrine' AGM
Reliable, tried-and-tested variety with delicious, large, white-fleshed fruit ready in late summer. A good garden cultivar.

'Rochester' AGM
The best all-rounder for under cover and outdoors, with firm, yellow flesh and crimson-flushed skin.

VEGETABLE

1 Make a Hardy Herb Container

(early April)

NOW THAT temperatures are warming up, it's an excellent time to create a container dedicated to herbs. Place it close to the kitchen in a sunny position and stroke the aromatic foliage regularly – it is wonderfully sensual. Use rustic wicker or gunmetal-grey galvanized metal: both suit the fresh, spring-zing of aromatically pungent herbs like lavender, sage, chives, basil, mint and thyme. Opt for a moisture-retentive, soil-based compost that's not too rich in nutrients. John Innes No. 1 is ideal: it will not dry out too readily.

A 60cm (2ft) wide container, usually circular or square, can be packed with thirty or more herbs, or you can plant a small wicker basket instead. The trick is to mix the textures and go for lower-growing varieties. Seek out some fine-leaved thymes. Add a prostrate rosemary to flow over the edge. Then select some larger, rounder leaves, including basil and sage. Both come in a variety of colours. 'Icterine' is a variegated golden sage, 'Purpurea' is a faded-damson and 'Tricolor' is a bold combination of green, purple and cream. Thymes and mints are equally diverse and some are lemon-scented whilst others are highly pungent. Go for crinkly parsley and lots of marjoram. Unite the planting with potfuls of upright chives randomly used throughout – then your container will satisfy the palate and look stylish. Order from an organic herb specialist (like Jekka McVicar), as the range and quality will be better.

Did you know? Aromatic herbs like sage and marjoram produce flowers that are very attractive to bees. Their nectar is highly concentrated. The most concentrated of all is the nectar of marjoram, *Origanum vulgare*, which contains 76 per cent sugar. August-flying butterflies adore it too.

SECRETS OF SUCCESS

- Pack the plants in tightly – this will look better and conserve moisture. Put the lowest-growing plants (like thymes) at the front.
- Water regularly for the first month until the plants are established, then cut down on the watering.
- Take cuttings of sages, thymes and oregano.
- Snip leaves and stems with scissors when harvesting.
- Remove and either pot up or compost the plants at the end of the year and clean the container ready for next year.

VARIETIES

Narrow-leaved Sage (*Salvia lavandulifolia*)
Small, narrow leaves and a neat habit make this the ideal sage for mixed planting.

Broad-leaved Thyme (*Thymus pulegioides*)
Strongly flavoured, dark leaves and pink flowers. Less twiggy than many thymes, so easier to prepare.

Garlic Chives (*Allium tuberosum*)
A late-flowering, white-flowered chive with a gentle growth habit. Mild garlic–onion flavour.

French Tarragon (*Artemisia dracunculus*)
The best herb for chicken, with narrow, linear leaves flavoured with aniseed. Take cuttings, as it hates wet winters, but it is hardier than most books imply.

Organic Tip ✔

Keep your container looking fresh by snipping back thymes at least twice a year so that they form a tight carpet. The other herbs should be cut back hard in spring, as most come from the Mediterranean, and not in autumn.

2 Sow Sweetcorn

(early April)

SWEETCORN used to be a trial to grow in years gone by and results were always poor. However, the modern F1 varieties are bred for cooler climates and will perform well if sown under cover now. Their hybrid nature makes them more vigorous and willing to crop. It also improves germination, so do opt for one of these.

Varieties are divided according to sweetness, with some listed as supersweet and some tendersweet. Each plant should produce two cobs, but once they are harvested the sugars quickly turn to starch. Make sure that your home-grown sweetcorn goes straight from plot to plate: then it will be much better than cobs bought from the supermarket. Supersweet varieties keep their sweetness for longer.

Sweetcorn has brittle roots and great care must be taken when handling plants. It's best to raise individual modules or pots to limit root disturbance. The seeds are large enough to handle individually. Almost fill the pots or modular trays with seed compost. Place one large seed in each, about 2cm (1in) deep. Water well, keep the seeds warm and they should germinate within 2 weeks. When the individual plants are 10–15cm (4–6in) high, harden them off for a week by putting your pots or trays outside. This will toughen up the foliage so that it is less attractive to slugs. Carefully plant them outside in blocks from mid-May onwards. These wind-pollinated plants need to be close together in order to produce cobs, so don't plant them in a row.

Did you know? The Iroquois tribe gave the first recorded sweetcorn ('Papoon') to European settlers in 1779. They had developed a 'three sisters' system of growing squash, beans and maize together on a mound – an early example of companion planting. The beans fixed nitrogen into the soil, making it more fertile, and they used the maize for supports. The squashes covered the ground, keeping the moisture in and preventing weeds from germinating.

SECRETS OF SUCCESS

- Choose a sunny, warm position that isn't too windy.
- Plant carefully, leaving 45cm (18in) between plants, preferably on fertile ground.
- Plant outside only once the fear of frost has passed, after mid-May.
- Water thoroughly throughout the growing season to encourage large cobs.
- Tap the upper stems to spread the pollen.
- Harvest once the beard (the hairy bit at the top) is brown. If in doubt, squeeze a kernel: if a milky liquid oozes out, it's ripe.

Organic Tip ✔

Don't put your sweetcorn outside too early in the year: this is a warm-season crop that is checked by cold spring nights. Planting through black plastic helps to conserve moisture and warm the soil, leading to improved yields.

VARIETIES

'Lark' F1
A pleasure to eat, this tender, mid-season variety produces thin-skinned, golden kernels even in the coldest gardens. A consistently high performer.

'Lapwing' F1
A mid-season variety known for producing the largest tendersweet cobs.

'Swift' F1
An early variety with golden kernels. if you're growing two kinds, make this one of them.

'Marei Yellow' F1
Vigorous, even in poor soils, with fully packed, bright-yellow cobs full of flavour. This one can be eaten raw or cooked.

3 Sow Winter Brassicas
(mid-April)

IT'S TIME to sow brassicas for winter use and this includes purple sprouting broccoli, kale and Brussels sprouts. You can create a seed bed outside and then plant, but it's far easier to raise individual young brassica plants under glass and then harden them off and plant them out. The most efficient way is to use modular trays large enough to accommodate a 7.5cm (3in) high plant. Use just one seed per module, because brassicas are good at germinating. They appear within 2 weeks and very few seeds fail.

Grow the plants on until they are between 10cm (4in) and 12.5cm (5in) in height (this usually takes 5 weeks), then harden them off for a week by putting them outside. Protect them from pigeon attack with wire. If they get short of food and water at this stage it will hamper their development for ever. Plant them into firm, but fertile ground as soon as possible. A dusting of blood, fish and bone (see page 317) helps to supply extra nitrogen. If the weather is dry, water the plants in well.

Brussels sprouts need the most space, with 60 cm (2ft) between plants. This allows enough light for the buttons to develop properly. Kale and purple sprouting broccoli can be spaced 45cm (18in) apart. Net with small-mesh butterfly netting straight after planting to prevent Small and Large Cabbage White butterflies from laying any eggs.

Organic Tip ✓

Social wasps are the most efficient of predators. They will tackle fully grown Cabbage White caterpillars by carving them into pieces. If caterpillars do infiltrate under your netting, remove it and allow predators a couple of days to attack. Do not use wasp traps: these creatures are meat-eating bees.

SECRETS OF SUCCESS

- Sow in April in modular trays.
- Get your plants in the ground before their growth spurt stops. Don't leave them lingering in trays to get stunted and starved.
- Fertile, nitrogen-rich soil is vital for brassicas. For this reason, many gardeners grow cabbages after legumes (beans and peas) as the roots of legumes fix extra nitrogen in the soil.
- If you haven't enriched the soil, add pelleted or powdered chicken manure or blood, fish and bone at 60g (2½oz) per square metre (10–11 square feet) before planting out.
- Wild brassicas tend to grow on sandy soil near the coast, so their cultivated cousins can cope surprisingly well in warm, dry summers.
- Netting brassicas with proper butterfly netting capable of keeping out white butterflies is vital, as this is the only organic way to prevent caterpillars from eating the leaves.

Did you know? In 1773 it became a criminal offence to steal or damage cabbages.

VARIETIES

Kale
'Cavolo Nero' (Black Tuscan Kale)
This hardy kale has long, linear leaves that can be picked in autumn long before the others.

Brussels Sprout
'Bosworth' F1
A mid-season sprout that can be picked before Christmas, although it stands well into winter. Not too tall and with a good flavour.

4 Sow Beetroot, Spinach, etc.
(mid-April)

FIND SOME space to sow some edible and ornamental members of the beet family now that the soil is finally warming up. These include beetroot, spinach, Swiss chard, sea kale beet and leaf beet. All five will need to be kept well watered in their early stages to prevent bolting. Apart from this, they are all easy to grow.

Beetroot, which was grown by the Ancient Egyptians on the banks of the Nile, enjoys warm conditions and moisture. However, 'Boltardy' AGM is more tolerant of extremes of temperature, so it can be sown before any others. It produces round, smooth-skinned, dark roots. The F1 hybrid 'Alto' is also excellent and its long, cylindrical roots push above the soil surface, clearly showing the size of the root. Earth them up if slugs are a problem.

Beetroot seeds resemble small sputniks and each one produces several seedlings – so sow thinly. You can use a narrow or a wide drill and the leafy thinnings can be eaten. There are monogerm varieties that produce one plant per seed, but cropping can be poor.

Leaf beet (which comes in vivid colours) has edible stems and leaves, and the white-stemmed sea kale beet is extremely hardy. Both should overwinter if sown between April and June. Spinach can be sown in any gap up to late August to produce a quick crop. Good AGM varieties include 'Scenic' and 'Toscane' for early sowing and 'Triathlon' and 'Spokane' for late sowing.

SECRETS OF SUCCESS

- Try to emulate the conditions on the Nile (as far as possible) with warmth and moisture. Always sow in warm, moist weather. If your garden is dry, plant 'Boltardy' beetroot. This bolt-resistant variety is capable of surviving dry, cold springs.
- Choose a warm, sunny position.
- Spinach is a great gap filler and can be sown until mid-August (perhaps later) and still crop within 10 weeks.
- Repeat-sow beetroot until July, as needed.

VARIETIES

Beetroot
'Alto' F1 AGM
The sausage-shaped, tender, sweet beetroot for slicing. This longer variety pushes upwards as it develops so you can see when to harvest. Produces an early crop.

Beetroot
'Boltardy' AGM
The spherical red beetroot for succession sowing – the most popular variety of all and deservedly so.

Spinach
'Tetona' F1 AGM
A heavy-cropping spinach with smooth, green leaves. Slow to bolt, and ideal for baby leaves or for cooking. Excellent downy-mildew resistance, so good for spring and autumn sowing.

Swiss Chard
'Bright Lights'
Colourful stems in red, yellow and orange, topped by red-tinted, edible leaves make this a dual-purpose vegetable, although stems and leaves should be cooked separately.

Did you know? The Romans used the dark roots of beet medicinally to cure fevers and restore good health. In the sixteenth century, the British name for beetroot was Roman beet, which suggests that the vegetable arrived with the Romans about 2,000 years ago.

Sow beetroot thinly and thin out if needed once the seedlings get to 7.5cm (3in). Give them space to fill out. A crowded row never works. When pulling, take beetroot from all along the row to allow those left to fatten up.

5 Apply Slug Nematodes
(late April)

TACKLE the slug and snail problem proactively now by watering on nematodes (tiny parasitic worms) in your key areas: your runner-bean patch or your hosta bed, for instance. Aim for a damp day and apply the nematodes after 4 p.m. to allow them the maximum chance of penetrating the soil effectively. One application in late April or early May (on damp soil) will solve your slug problem for 6 weeks or more – but most die underground, so the evidence will be scant. The nematodes are easy to apply: simply use a watering can with a coarse rose.

Place small empty flowerpots on their side in your vegetable patch and you will often find snails resting inside. Destroy them at dusk (with a big boot on a path) and often slugs will feast on their remains – allowing you a second offensive. Hoe your vegetable patch thoroughly to disturb the soil: this brings eggs to the surface for thrushes, blackbirds and robins.

Protect your key plants with 'decoys': the best slug magnets are lettuces and African marigolds, both of which attract slugs at dusk. Collect and destroy them – wear rubber gloves if squeamish. Always harden off young plants to toughen the growth and avoid quick-release plant foods. These promote soft, sappy growth – a gourmet feast for gastropods.

SECRETS OF SUCCESS

- Certain plants attract slugs more than others. Protect vulnerable crops, like runner beans, with decoy plants and then collect the slugs off the decoy plants.
- Slugs prefer soft, leafy growth. Don't over-feed your plants with nitrogen (it will make them irresistible) and always harden them off for a week when they come out of the greenhouse or from a garden centre.
- Aromatic plants and silver-leaved plants are unpopular with slugs.

TYPES OF SLUG

Grey Field Slug (*Derocereas reticulatum*)
The most common and serious slug pest. Usually light grey or fawn and measuring 3cm (1½in), this is the very soft-bodied slug you find in lettuces and cabbages.

Keeled Slug (*Tandonia budapestensis*)
Grey-black with a ridge down the back. These are larger than the Grey Field slug (about 6–7cm/2½in) and they tend to live and feed underground. Potatoes are a delicacy to them.

Black Slug (*Arion ater*)
This is the large slug you are likely to see in the daytime after rain. It can measure up to 20cm (8in) and is black in colour, although some subspecies have a distinctive orange colour. This type cleans up debris – don't kill it. It feeds on rotting foliage, fungi and petals.

Garden Slug (*Arion hortensis*)
A slug with tough, leathery skin. Darker in colour (grey–black) with a paler, yellowish underside. Destructive at every level – and this one can climb.

Organic Tip ✔

Many songbirds, hedgehogs, beetles, frogs and toads will eat your slugs and snails and their eggs, so it isn't a good idea to use slug bait, as the chemicals may get into the food chain. Bait also kills the debris-eating slugs that clean up the garden. The copper compound used to colour the pellets blue is highly toxic to frogs and toads. Tests have shown that ground beetles find poisoned slugs more attractive to eat, with catastrophic results for natural slug control.

Did you know? Not all slugs are bad for the garden. Many of the larger ones clean up plant debris under the ground, so they shouldn't be killed indiscriminately. The time to catch the real baddies is half an hour before dusk, when they are at their most active.

6 Sow Cucurbits
(late April)

CUCURBITS – cucumbers, courgettes, pumpkins and squashes – are the most frost-tender plants in the vegetable garden. You can sow them under glass now, but these plants should never go out into the garden until the first week of June. If you put them outside in May, be prepared to cover them at night. The slightest frost will turn them to mush and cold nights will check their growth so severely that they will never recover. They resent cool temperatures and wind bruises the leaves very easily.

Varieties differ a great deal. If you raise your own plants you can select good varieties suited to your needs. The seeds are large and oval and they can rot if laid flat on to compost. Place them vertically to avoid this problem, pushing them down about 1cm (½in). There is no reason to chip or soak seeds – they generally germinate well in warmth and light.

Place one seed in a 7.5cm (3in) pot almost full of seed compost. Label clearly, as cucurbits all look very similar when young. Water thoroughly, and then leave the pots until the seeds have germinated. It will take several weeks for a plant to develop to the planting-out stage and it's best to keep watering the pots every morning so that the seedlings sit in drier compost overnight. Fleece the young plants if a cold night is forecast.

These days the marrow has been usurped by the winter squash, which has to be stored for at least 6 weeks to develop its nutty, sweet flavour. There are excellent outdoor F1 hybrid cucumbers and many courgettes that produce small fruits that won't turn into marrows.

SECRETS OF SUCCESS

- Grow the plants to a good size under glass before planting out. Feed with liquid nitrogen-rich plant food if necessary.
- Fleece if a frost is forecast.
- Water well during summer and feed with chicken manure, which is available as a powder and as pellets. Ease off the watering in autumn.
- The grey-skinned 'Crown Prince' and 'Blue Hubbard' squashes keep the longest – often until late April. However, they are thin-skinned and more susceptible to frost damage once harvested.

Did you know? Pumpkin seeds have been found in burial caves in Mexico dating back to 8000 BC. Pumpkin is the food that sustained the Pilgrim Fathers and in October 1621 at the first Thanksgiving Day meal boiled pumpkin was served, but these days Americans prefer pumpkin pie instead.

VARIETIES

Courgette
'Romanesco' AGM
Long-cropping, this Italian variety produces slender, ridged courgettes with a nutty flavour.

Courgette
'Venus' F1
A compact courgette producing shiny, dark-green fruits. Could be grown in a container.

Cucumber
'Euphya' F1
A top-quality all-female variety that's productive, refreshing and has unbeatable long, straight fruit, easy for picking.

Cucumber
'Marketmore' AGM
The most prolific small cucumber for outdoor use. Spiny, tasty fruits and lots of them.

Winter Squash
'Sunshine'
A Japanese squash, round, orange and with a superb flavour. Each one weighs 2kg (4lb) and keeps until late January.

Winter Squash
'Uchiki Kuri'
The Japanese red onion squash, which is almost pear-shaped. The flavour is excellent and this also keeps until late January. 2.5kg (5lb).

Organic Tip ✔

Once your cucurbits go outside, cloche them at night using rigid plastic or glass bell cloches. This will double their size and produce an earlier crop, because they need warmth to photosynthesize.

MAY

May is a glorious month and it marks the transition between spring and summer. Much of the hard work in the vegetable garden has already been completed: most hardy crops are in the ground and hopefully racing away and full of promise. You can almost hear things growing, but this includes the weeds, and it's important to get on top of them. Make use of the garden hoe, the finest implement ever created. Use it every week, between the rows, to get rid of weed seedlings. It will also disrupt slugs, the enemy of every gardener, because slugs hate soil disturbance. Hoeing also aerates the top layer of soil, helping rain to penetrate, but its usefulness doesn't end there. The fine dust you create acts as a mulch and helps to keep moisture in the soil.

The difference between night-time and daytime temperatures can be at its greatest in May and these see-sawing temperatures create poor growing conditions, especially for frost-tender plants. Don't be in a hurry to put any tender crops outside until late May or, better still, early June.

In the fruit garden crops will always be better if May is warm and sunny. However, if it's a dry spring you may have to water your fruit because drought conditions at this stage, when fruit is setting, can be disastrous. Try to water in the first half of the day, when temperatures are mild, rather than in the evening. Pears, strawberries and blackcurrants are really thirsty plants, so give them priority. Try to keep water away from the foliage because splashes could spread fungal diseases. If a late frost is forecast, be prepared to cover blossom with fleece or newspaper if necessary. Strawberries are particularly vulnerable and, once frost blackens the centres of the flowers, you won't get any at all.

FRUIT

1 Straw and Net Strawberries

(early May)

GROWING strawberries is very worthwhile, but you won't get a crop unless you protect them from the three main enemies: the rain, the slug and the blackbird.

The best way to protect fruit from rain damage is to cushion them on a bed of straw and this is best applied now on a dry, fine day. Simply lift up the foliage and the forming fruit trusses and lightly arrange the straw underneath. If the bale is very tightly packed, open up the straw with your hands. Try to avoid musty bales. You need about half a bale for one 2.4 × 1.2m (8 × 4ft) bed.

Once the straw is down, netting can be done too. The easiest system relies on a set of short, sturdy stakes (about 60cm/2ft high). Arrange these to form a neat grid with roughly 60cm (2ft) between each stake. Pop a small plant pot over each stake – terracotta ones look far more decorative than plastic. Then place black plastic netting over the pots so that the netting is held above the fruit: this will repel aerial invasion. Thread long bamboo canes along each edge so that these rest on the ground to form a mini fruit cage that you can lift on and off in seconds. The canes will also prevent birds from creeping underneath.

When your protection system is in position, keep an eye on your fruit in case botrytis strikes. If you see any grey fluffy mould on the berries, cut them off and bin them. Check over the rest of the crop.

Strawberry plants form the next year's flower buds after they have finished fruiting for the year, a process that is light-dependent.

So, once the crop has finished, remove all the straw. Then cut away all the leaves close to the crowns, being careful not to damage them, and feed with a potash-rich fertilizer such as comfrey tea (see page 326). Your plants will have fresh foliage within 10 days. Cutting back also helps to clean the bed of aphids and this will help to prevent virus.

For advice on dealing with slugs, see page 103.

Did you know? The name 'strawberry' is derived not from straw but from an Old English word meaning 'to stray', because of their habit of producing runners and straying off.

Organic Tip ✔

If you are going to water strawberries, always do so in the first part of the day because slugs become active at dusk and like to glide over wet surfaces. Watering in the evening encourages trouble.

SECRETS OF SUCCESS

- Be vigilant in watching out for mould. Remove any affected fruit as soon as you see it.
- Pick ripe fruit regularly – aim for every other day.
- Always remove any strawberry plants with mottled, ringed or yellow-veined foliage – these are symptoms of virus.

VARIETIES

For varieties of strawberry, see February, page 42, and March, page 72.

2 Tend Pears and All Wall-trained Fruit

(mid-May)

WALL-TRAINED fruit is usually positioned against sunny walls and it is always going to be short of water in May, even when the weather is wet, because walls shelter the ground from the rain and then act like a wick in dry spells, drawing water from the soil. Try to get into a regime of regular watering, especially with young trees, which will not have developed deep roots. Trees that are decades old may have a root system that's capable of accessing water held deeper in the ground.

It's well worth making the effort to water well, for if trees become drought-stressed they shed their fruit early. If you can nurse them through May and June far less will drop off. Pears in particular really love moisture, so water them from the moment the flower buds burst until 6 weeks after blossoming to improve the yield. Gently tip one bucket of water on each pear tree every day in dry spells.

Most trees shed fruit in June as a natural process – it's known as the June drop. However, if the fruit set is very heavy it is worth thinning at the end of June–early July; see page 146. Plums can be so laden that the branches snap and many old trees are propped up on wooden supports; this could be an option with trees in open ground.

The most common shape for wall-training is the fan, which radiates from the base of the tree. This must be kept in shape and any growth that is straying away from the wall should be nipped back now. Either rub it off with your fingers or snip it, sterilizing the secateurs after every tree. Dipping them in boiling water is my system, but some people use bleach. Keep the growth tight against the wall throughout summer.

For advice on pruning and shaping wall-trained fruit, see page 232.

Did you know? Training exotic fruit trees (like pomegranates, oranges, peaches and figs) began in France, but by the seventeenth century it had become widely practised in the Netherlands. Most of the early fruit trees grown by British pioneers like John Tradescant the Elder were imported from Holland. When William of Orange was crowned King of England in 1689 the two countries forged much closer links and soon fruit-training became popular in England too. Many walled gardens had their walls raised and buttressed to accommodate wall-trained fruit.

SECRETS OF SUCCESS WITH WALL-TRAINED FRUIT

- Buy top-quality fan-trained or cordon-trained trees from a reputable fruit specialist. Try to go and choose your tree personally. They are expensive, but buying one of these will save you years when compared with buying and training small one-year-old trees (or 'maidens', as they are known).
- Add supports before or when planting.
- Spend time keeping your trees in shape – see page 232.
- Water your trees throughout the growing season for the first few years of their life.
- If disease strikes, tidy up any diseased material thoroughly.
- Be prepared to thin your fruit.

Organic Tip ✓

Peach trees are quite hardy, but they are usually tucked up by walls because they flower so early in the year that the blossom gets frosted. Try to protect them (see page 306) and always do the pollinating by hand (see page 77) because so few bees will be about that early.

FAN-TRAINED VARIETIES FOR A WARM WALL

Apricot
'Flavorcot'
Specially bred for the cooler UK climate. Large, egg-sized, orange-red fruits with excellent flavour. Late-flowering and frost-resistant, so you should always get a crop. Harvest in August. Look for trees grafted on to Mont Clare rootstock.

Apricot
'Tomcot'
An early-cropping apricot giving a very heavy crop of large, crimson-flushed fruits which are ripe for picking from around the middle of July. Look for trees grafted on to Mont Clare, a dwarfing rootstock, as these are suitable for training. (Torinel rootstock is best for full-sized trees – see page 78.)

Cherry
'Sunburst'
This very dark cherry has almost black fruit with a rich flavour. Net against birds in May and harvest in early July. Self-fertile. Pollination Group D.

Nectarine
'Snowqueen'
This new French variety crops heavily. It ripens in late July and is considered to have the best nectarine flavour. Protect from winter rainfall to control leaf curl. Can also be grown under glass. Look for trees grafted on to Mont Clare rootstock.

3 Plant Blueberries
(mid-May)

BLUEBERRIES have acquired super-food status in recent years because they are packed with antioxidants. They produce a useful crop over 4–6 weeks in late July and August (following on from strawberries), when soft fruit is generally in short supply.

It is quite possible to grow them, but there is one major drawback. Blueberries are members of the heather family (*Ericaceae*) and they grow wild in the heaths and forests of temperate North America, which means that they enjoy a cool, moist root run in well-drained, humus-rich, acid (pH 4.5–5.5) soil. You have to give them these conditions in order to grow them successfully in the garden, which means growing them in containers for most of us.

Wooden containers tend to be better for blueberries than terracotta ones: the latter get warm and absorb heat. Place a plastic sheet in the bottom of the planter and run it some 10cm (4in) up the sides. Cut a few holes in the base to ensure slow but steady drainage. Add ericaceous compost, plant and then mulch well with leaf mould, if you can, to help keep the root run cool and moist. Water with rainwater only.

Blueberries are long-lived plants. They fruit on branches and side shoots that were produced in the previous year. Prune in winter when the plant is dormant, removing the old, tired woody growth that has fruited for 2 years or more and any other weak or damaged stems. Take out any branches that are crossing or congested in the centre of the bush. Young bushes are not normally pruned for the first 3 years.

The varieties on offer (bred from *Vaccinium corymbosum* – the northern highbush blueberry) are subdivided into early-, mid- and late-season kinds and, although blueberries are partly self-fertile, you get a better crop when cross-pollination takes place. Ideally you want two varieties from the same group, and mid-season and late varieties are the most useful. Mine are usually pollinated by a small rusty-backed bee, *Bombus pascuorum* (the Brown-banded Carder Bee), so it's important that they are placed in sunlight to encourage the nectar.

Did you know? Blueberries are recent arrivals in Britain. The first plants were donated by Canada in 1951 to help economic recovery following the Second World War. They came from Lulu Island, just south of Vancouver, and were accepted by the Trehane Nursery, based on acid heathland at Wimborne in Dorset, which went on to become the first to grow blueberries commercially in Britain. They began selling them in 1957. Amazingly, some of the original bushes still survive.

SECRETS OF SUCCESS

- Blueberries do not require heavy feeding. In April, lightly sprinkle a granular fertilizer suitable for acid-loving plants (e.g. Vitax conifer and shrub fertilizer) over the soil surface. Alternatively, a liquid feed (e.g. Miracid) will do just as well.
- If you garden organically, avoid any fertilizers that might contain lime, such as manure, chicken manure and blood, fish and bone. Use hoof and horn instead, or ericaceous fertilizer.
- Water bushes well, but don't allow them to become waterlogged. Use at least 4.5 litres (1 gallon) of rainwater per container every week in summer. Tapwater is too alkaline for blueberries.
- Protect them from excessive winter wet by moving them close to a wall. They are very hardy, but they resent having cold, wet roots.
- Marauding blackbirds love to strip the fruit. Net your plants or put them in a fruit cage.

Organic Tip ✔

Mulching helps blueberries to stay moist and cool in summer. You can use bracken or leaves of trees such as oak and beech, all of which will keep the pH on the acid side and maintain the humus content, as well as adding some fertility.

VARIETIES

'Bluecrop'
The most widely grown variety. Tall, producing a heavy crop of large, light-blue berries with a good flavour. Mid-season.

'Chandler'
A bushy variety that produces much larger fruit – about twice the size of most blueberries. Mid–late-season.

'Patriot'
A rugged variety recommended for heavier, wetter soils and colder areas. Produces large, slightly flat, velvety berries with firm flesh and an excellent flavour. Mid-season fruit, but can flower early.

'Herbert'
Heavy, compact clusters of very large, dark-blue, highly flavoured fruit on a low bush with a spreading habit. Mid-season. Superb autumn colour.

4 Try Some Melons
(mid-May)

MELONS are a challenge for any gardener and success can never be assured. However, if you are a keen fruit-grower with a cold frame in a sunny position or a greenhouse, this is the time to think about acquiring some plants or sowing some seeds.

There are three types of melon. Honeydew varieties have firm yellow flesh. Musk melons have yellow- or green-netted skin and they are the most difficult to grow, even under glass. Cantaloupe melons, which have rough skin and orange-tinted flesh, are the easiest group to grow.

Warmth and water are the keys. All members of the cucurbit family will photosynthesize only in warm temperatures, so it's vital to boost the warmth early in their lives.

Melon plants can normally be acquired in May, but sowing your own now will also work, even though it is a little late (April is best). Melon seed germination is good, so put one seed, pointed end down, in a 7.5cm (3in) watered pot of airy seed-sowing compost and cover with a thin layer of vermiculite. When the plant has three or four true leaves, plant in a cold frame or greenhouse, spacing them 90cm (3ft) apart. If planting in a frame, harden them off by putting them somewhere sheltered during the day for at least a week before planting.

Melons have to be trained and pruned so that they bear good-quality fruit and stay within their allotted space. Pinch out the growing tip at the fifth leaf. As side shoots form, pinch out all but the strongest four and train them to occupy the desired space. Pinch out again; secondary shoots, on which the flowers form, will then grow; pinch the secondary shoots out two leaves beyond the developing fruit.

Organic Tip ✔

Open the greenhouse or cold frame during the day when the melons are in flower to allow pollinating insects to visit. The drier air helps pollination too.

Did you know? Melons produce male and female flowers. The male flowers (held on longer stems) are produced first and then female flowers (each with a small, bulbous baby swelling at the base) follow. Pollen has to move from the male to the female: the cluster of stamens in a male flower is large enough to break off and hold between your finger and thumb – just dab it on to the stigma of the female flowers.

SECRETS OF SUCCESS

- Choose a reliable Cantaloupe variety.
- Give melons a warm position that gets lots of sun. However, greenhouse melons benefit from being shaded, as the intense heat under glass can scorch them.
- Water your melons often and feed them with a tomato fertilizer every 2 weeks to encourage strong growth and more flowers.
- Plant in a raised bed – these warm up faster. Or use a growbag, two plants per bag. Bring your bags inside for 3 weeks before planting up so that the contents warm up.
- Train the plants vertically, and once the fruit is grapefruit-sized, use onion netting (or something similar) to support it.
- Fruit quality suffers if you try to ripen too many fruits on each plant, so restrict the number of fruit. Cold-frame melons should bear only four fruits. Greenhouse melons can produce six.
- Melons develop an aroma when they are ripe, so use your nose to decide when to harvest them. They also get a woody stem that starts to split.

VARIETIES OF CANTELOUPE

'Ogen'
A light-green, Israeli variety with a rich, aromatic sweetness. Crops in 90 days.

'Sweetheart' F1 AGM
A high-yielding melon with medium-sized, grey-green fruits and light-orange flesh. One of the quickest to mature.

'Fastbreak' F1
Good-sized fruit with succulent, sweet orange flesh. Ideal for the greenhouse, cold frame or cloches, but can also be grown outdoors in some areas of the country.

5 Install a Codling Moth Trap
(late May)

MANY OF you will have bitten into an apple and found an annoying little dark-headed white caterpillar in the middle – or worse still, half a white caterpillar. These are the maggot-like grubs of the inconspicuous grey and brown Codling moth (*Cydia pomonella*). The adult Codling moths – which measure about 1cm (0.4in) – emerge in late May and early June and lay their eggs on or near developing fruits between June and mid-July. They fly and mate on warm, still evenings when the temperature is above 13°C (55°F). When evening temperatures reach 15.5°C (60°F), conditions are the most conducive to egg-laying. Eggs are laid singly on leaf or fruit surfaces; then the caterpillars hatch out and eat into the fruit. Once they have had their fill, they overwinter in leaf litter or under loose flakes of bark and they pupate the following spring. Codling moths can also attack quince and walnut, but these are rarely infested to any great extent.

In order to thwart these pests, commercial growers hang up Codling moth traps, which alert them to the presence of the moths and capture some of them. The size of a commercial orchard means that when numbers build up the growers still have to spray with insecticide, but they need less because the traps have disrupted the breeding cycle. The home gardener, however, with only a couple of apple or pear trees, may solve the problem entirely by just using a trap. The open-sided boxes are available from garden suppliers and need to be hung in the tree in May. Each one contains a pellet that emits female pheromones at a constant rate, attracting male moths into the trap. The sticky sheet on the bottom traps the male when he homes in on the pellet, hopefully before he has found a female to breed with.

Did you know? Codling moths are the most serious problem in apple production throughout the world. In parts of America (such as Utah) the moths produce five generations per year. In Britain we can get two generations, but only if the first is early enough to be at the cocoon stage by the beginning of July.

Organic Tip ✓

Birds and wasps feast on small grubs, so encourage them to stay in your garden. Feed the birds in winter and do not attack wasps: they are the most effective predators of caterpillars in many gardens.

SECRETS OF SUCCESS

- Five traps are recommended for a 5-acre orchard and the advice is to hang one at the edge with the rest in the middle. One trap should be enough for most gardens, however.
- Hang your traps in the upper third of the tree.
- Check your traps every day until you find the first moth and then you will know they are on the wing.
- Traps can be stored in early August and refills (sticky sheets and pheromone pellets) can be bought for subsequent years.

6 Thin Fruit
(from late May)

WHEN WE look at a tree laden with fruit, a pleasant glow comes over us. However, we should not let quantity overwhelm us: quality is much more desirable. We do not want the tree or bush to become so exhausted that it rests for 2 or 3 years. With this in mind, it's best to thin out your fruit so that the tree or bush produces a top-grade crop. Depending on the type of fruit, thinning should take place from late May, throughout June.

Gooseberries often yield a bumper crop of small fruits. Pick every other one towards the end of May and cook them, then leave the rest to grow larger. If you're planning to allow your gooseberries to colour up and become really ripe, net the bush after thinning.

Nectarines, plums, peaches, damsons and gages should be thinned twice – once when the fruit is the size of a hazelnut and then again in late May when the fruit is about 2.5cm (1in) in length or golf-ball size. Try to create a 5–7.5cm (2–3in) gap between individual fruits by snipping some of them away with scissors. Do not pull them off: you may destroy next year's fruit buds. Peaches growing along the same branch should be 22cm (9in) apart.

With apples, the number may depend on the tree. Many old apple trees with triploid blood (like the magnificent 'Bramley Seedling') bear a heavy crop every other year. These trees are probably best left to their own devices. All apples and pears will shed fruit in June in a natural thinning process (the June drop) and for this reason apple- and pear-thinning is best left until mid-June. Some varieties shed their fruit later and thinning can take place right up to mid-July and still be effective. Final thinnings of dessert apples should leave 10–15cm (4–6in) between fruit. Large cooking apples often need to be spaced to 15–22cm (6–9in).

Pears need less thinning than apples and the time to start is after the June drop. However, you must wait for pears to turn downwards before thinning them, so that you can be sure that the fruit has set. Usually pears are thinned to two per cluster. If the branch looks weak, prop it up using a wooden stake with a forked head. Try to cushion the area between the branch and the stake with soft material or rubber – sections of tyre work well.

Thinning grapes is also a June pastime. You need to shorten the trusses but leave the shoulders of the bunches intact. Cut up into the bunch, removing most of the interior grapes and any very small ones. There are special long-bladed scissors for this and generally two pairs are used. The technique is not to touch the fruit, but only the stems. After thinning, inspect the fruit every week, cutting out any split or diseased grapes.

Did you know? The earliest descriptions of fruit-thinning appear in 1771 in a book about vines. The process was carried out with great precision. A long, pencil-like stick was used to separate the bunches of grapes and the gardener would then snip and shape the bunches. The thinnings were taken to the kitchen to be made into tarts and pies, or they were juiced.

Organic Tip ✔

When thinning apples, look for the 'king apples' at the centre of each cluster. These are generally misshapen and often have no stalk. Remove them before thinning the rest of the cluster, as getting rid of this 'cuckoo' produces bigger fruit.

VEGETABLE

1 Plant Tomatoes under Glass

(early May)

IF YOU have an unheated greenhouse it's time to put in your tomato plants. There will be plenty on offer in garden centres. Look for good-sized plants that have green foliage. If the foliage looks even slightly blackened, don't buy them: these plants have suffered a cold shock and it is likely to hamper them permanently.

Some varieties are best trained as cordons, others left to grow as a bush, and some have a trailing habit. Varieties are therefore classed as one of these forms: cordons, bushes and trailing. Cordons are easier to manage because they take up less space, but they do need staking securely. The side shoots are pinched out to create an upright plant and fruiting trusses are limited to six. Bush varieties and trailing (or tumbling) varieties are allowed to develop naturally. Their habit varies, but some sprawl and take up a lot of ground room. They are suited to growbags and containers. Fruit varies in shape from plum to cherry, from beefsteak to the more conventional round fruit. There are reds, oranges and yellows too, although I have yet to find a yellow variety with a good flavour.

Recently, grafted tomatoes (which have been used by commercial growers for decades) have become available to the home gardener too. These have the advantage of being guaranteed to be free of soil-borne disease. They are grafted on to vigorous rootstocks and this makes them grow far more quickly and aggressively. Be prepared to remove some of the leaves if necessary so that the fruit is exposed to the sun. They can make huge plants that threaten to take over your greenhouse. The ones I have tried ripened poorly when grown outdoors.

Did you know? Tomatoes belong to the Solanum family (just like potatoes) and they also suffer from potato blight, so don't grow the two close together. Both come from Peru, but wild tomatoes thrive in the warm, damp lowlands that separate the Andes from the Pacific while the potato is a high-altitude plant.

SECRETS OF SUCCESS

- Try to keep the temperature as level as possible. Make sure that the greenhouse is shut up well before dusk and that it is ventilated during the day. Ideally, temperatures should be between 20°C and 24°C (68° and 75°F).
- Plant in fertile soil, and stake and support immediately.
- Water in well when you plant, then stand back for a few days to encourage the roots to search for water.
- Once you see the plants beginning to grow, water them again. Stick to a regular watering regime: uneven watering causes the tomatoes to split. Try to water in the morning if possible.
- Pinch out the side shoots of cordons with your fingers. They appear in the axil between stem and leaf. Wear gloves if you have sensitive skin.
- Once the first truss of flowers is set, water on a high-potash commercial tomato food or diluted comfrey tea (see page 154) every 14 days.

Organic Tip ✔

Tear off tattered foliage, etc., rather than cutting it. Commercial growers adopt this approach as it prevents disease entering the plant.

VARIETIES

'Sungold' AGM
Large trusses of sweet, orange cherry tomatoes. Treat as cordon.

'Beefsteak' AGM
Very large, wide, fleshy fruit with a good flavour – cooks and eats well. Treat as cordon.

'Tigerella' AGM
Striped red-and-green fruit with a good flavour. Treat as cordon.

'Gardeners' Delight' AGM
Large trusses of red cherry tomatoes.

'Sweet Olive' AGM
Do not remove side shoots. Produces upright plants with cascading trusses of plum-shaped, shiny red fruits. Can also be grown outdoors.

'Cristal' F1
Treat as a cordon and remove side shoots. Produces very round, dark-red fruits with dark, ruby-red flesh. A high yielding, disease-resistant variety.

For outdoor varieties, see March, page 95, and August, page 245.

2 Sprinkle Orange Annuals
(early May)

VEGETABLE gardeners need their pollinators and bees are obviously vital. Hoverflies also pollinate and their almost transparent larvae predate aphids and other small creatures, providing pollination and pest control. Adult hoverflies lay their eggs near aphid colonies and then their maggot-like larvae feed on them.

Hoverflies need nectar for flight and pollen for breeding, but they have tiny mouths and therefore prefer tiny flowers. Umbellifers are highly attractive to them. The green-flowered *Bupleurum griffithii* 'Decor' is an easily grown annual and it will self-seed once established. *Ammi majus, A. visnaga* and *Orlaya grandiflora* are three other umbellifers with white flowers that could also be used. These need to be sown in trays and planted outside.

Hoverflies are also strongly attracted to bright orange and yellow, and the orange pot marigold (*Calendula officinalis*) will lure them in. This is one of the easiest annuals to grow. Simply sprinkle the crescent-shaped seeds on to damp ground and lightly cover them with soil or compost. Let them grow to flowering size, then dead-head them until late summer. After that they can be allowed to self-seed. French marigolds (*Tagetes patula*) are also excellent and plants should still be available at garden centres. Avoid the double forms: they are less insect friendly.

SECRETS OF SUCCESS

- Grow a range of annuals that flower between midsummer and late autumn. Try to avoid fully double-flowered forms, because the extra petals replace the stamens and other flower parts. Insects cannot access the flowers and, even if they could, there is rarely any pollen or nectar.
- Choose hardy, easy annuals that can be sown directly into the soil.
- Choose a warm, damp day and rake over the soil, sprinkle on the seeds and add a light covering of compost or soil.
- Dead-head regularly to prevent seeds forming until late August, then allow some plants to set seed.
- Save your own seeds. Collect them on dry days and packet them up. A biscuit tin in the fridge is ideal for storage.

Did you know? The Marmalade hoverfly is very common in gardens. It is one of 290 species of hoverfly in Britain. Its orange body has double black stripes. One larva will eat several hundred aphids during its 14-day lifespan.

VARIETIES

Cornflower
Centaurea cyanus
The blue cornflower is one of the few true blues of summer. Sown now, it is usually in flower by midsummer. This is the preferred annual flower of the red-tailed bumblebee, but hoverflies also visit.

Cosmos
***Cosmos bipinnatus* 'Antiquity'**
A shorter cosmos in a mix of pinks. This simple saucer flowers from late June until late into the year if you dead-head. Popular with bees and hoverflies.

Pot Marigold
***Calendula officinalis* 'Indian Prince'**
A classy, maroon-centred, single orange pot marigold. It is so much more rewarding for insects than the very full doubles often sold.

Scabious
Scabiosa atropurpurea
This will supply a succession of pincushion scabious throughout summer and autumn. Rich maroons, whites and Beaujolais mixes are all on offer and the insects seem to like them all.

Organic Tip ✔

Try to plant your annuals in a warm position, preferably where afternoon sun falls, because nectar always flows more freely in warmer temperatures. Leave flowering annuals in the ground as late into the year as possible for foraging bees.

3 Feed Your Plants with Comfrey Tea

(mid-May)

YOUR SUMMER bedding plants and vegetables in containers are probably growing in commercial compost which will sustain them for only 6 weeks at most. So now is the perfect time to start watering on a fortnightly liquid feed to boost flowers and fruit.

Choose your water-on plant food wisely. Some nitrogen-rich feeds only promote leafy growth. These are ideal for boosting young yew hedges, cabbages and container-grown topiary, but for flowers and fruit you need to use a potash-rich liquid tomato food.

You can make your own high-potash food for free using comfrey leaves. Comfrey is a member of the borage family and 'Bocking 14' is the best clone because it doesn't produce seeds. Put the chopped leaves in a container with a lid and leave them to rot for a couple of weeks. No water is required. As the leaves decompose they produce a brown liquid called comfrey tea. This can be diluted one part comfrey tea to twenty parts water.

The liquid can smell, but Garden Organic at Ryton produce a wall-mounted drainpipe system that saves you opening the lid. The pipe has removable lids on both ends. The leaves go in the top and, once rotted, the liquid is drained from the bottom.

Chopped comfrey leaves can also be used as an accelerator on the compost heap to speed up decomposition. Just make a layer of leaves on the top. Plant three or four plants of 'Bocking 14' by your compost heap so that it's handy.

Did you know? The benefits of growing comfrey were discovered by Henry Doubleday (1810–1902). He wanted to patent a glue for postage stamps and he imported comfrey from Russia with the aim of extracting glue from the mucilaginous roots. He discovered that comfrey stayed in leaf for 10 months of the year and wrote about his findings in the *Gardener's Chronicle and Agricultural Gazette*. Years later, the famous organic gardener Lawrence Hills (1911–91) discovered the articles and visited Bocking, the village near Braintree in Essex where Henry Doubleday had lived. He found Henry's descendants (nonagenarians Edith and Thomas) in residence – still growing comfrey. Lawrence Hills identified a strain which he called 'Bocking 14' as the best. Its leaves contain twice as much potash as ordinary comfrey.

COMFREYS

Symphytum uplandicum
'Bocking 14'
Originally imported from Russia, this kitchen-garden clone is available from organic gardening catalogues.

Symphytum caucasicum
This pretty comfrey is tall and elegant with sky-blue flowers and soft-green leaves. It is adored by bees, so makes a good addition in a wilder area of the plot.

Organic Tip ✔

Comfrey leaves should always be placed at the bottom of the potato trench because as they decompose they release nutrients that boost your crop.

SECRETS OF SUCCESS

- Comfrey produces nectar-rich flowers in May, just when other flowers are in short supply. Pulmonarias are in the same family and they too flower well in spring. Both are popular with bees. However, when you grow 'Bocking 14' for tea, don't let it flower.
- After flowering, cut the leaves from the base to leave a stump. Chop the leaves and stems up roughly and make the first batch of comfrey tea. Repeat the process throughout summer. You should be able to get at least four cuts.
- The ratio of nutrients in 'Bocking 14' is almost the same as in commercial tomato foods. Comfrey tea should be used every 2 weeks to promote flowers and fruit in the vegetable and flower garden. It will save you the expense of buying commercial tomato feed.
- Once you've planted comfrey, it's difficult to dig up. Always place the plants carefully and avoid any smothering groundcover varieties such as *Symphytum grandiflorum*. Remove the flowering stems as they fade.
- 'Bocking 14' is sterile, but other types self-seed prolifically.

4 Sow Runner and French Beans Outdoors

(*mid-May*)

IT'S TIME to sow frost-tender runner beans and French beans straight into the ground, because the soil is now warm enough to ensure rapid germination. Always put the poles up before sowing any climbing varieties. I find a wigwam of eight tall canes (securely tied at the top) withstands strong winds much better than a long row. Beans have soft leaves and they suffer horribly in strong winds. Sow three beans round each cane and thin if necessary once they come up. Also sow a handful in the middle of the tripod or at the end of the row for filling in gaps.

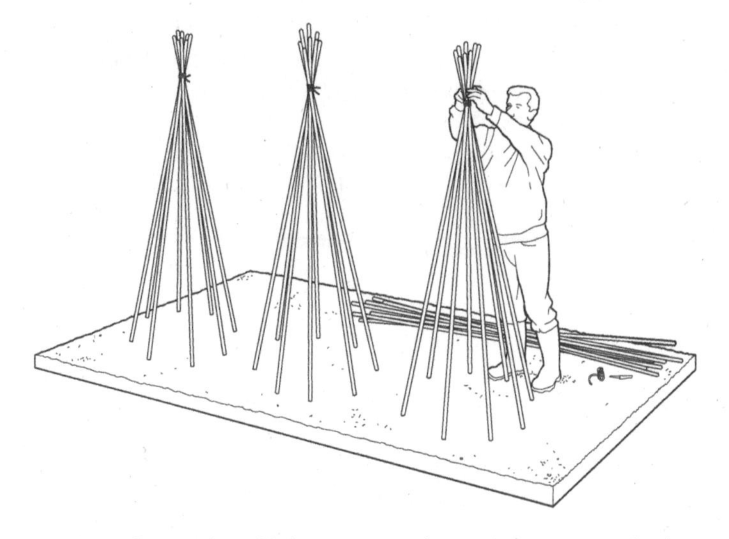

Keep slugs at bay. If they nip out the growing points the beans will not recover. Adding lettuces to the area helps to lure them away from the beans.

Try to sow at least two varieties of runner bean because they are influenced by the weather. White-flowered runner beans have paler seeds and red-flowered forms have darker seeds. As a general rule, the paler the seed the more heat-tolerant the variety is. Red-flowered beans often drop their flowers in hot weather once the night-time temperature reaches 16°C (62°F). White-flowered varieties thrive in warmer summers.

Climbing French beans also prefer hot summers and the dark-podded 'Blauhilde' and the green 'Cobra' are star performers. Flat-podded varieties (like 'Pantheon' and 'Hunter') crop very heavily whatever the weather. Growing a selection of varieties will ensure a crop no matter what summer brings. New hybrid varieties with both French and runner-bean blood are available.

Did you know? The runner bean was grown as an ornamental plant when it was first introduced in 1633 by John Tradescant the Elder. He became gardener to Charles I in 1630 and grew scarlet runners as an ornamental flower on arbours. The handsome flowers, which appear in August once the day length has shortened, became popular and were added to nosegays because they lasted longer than many others. These early introductions rarely set seed.

SECRETS OF SUCCESS

- Although beans and peas generally resent extra nitrogen, preferring to fix their own with their root nodules, runner beans seem to need it. The best way to provide lots of nitrogen is to incorporate organic matter either by double digging (see January, page 25) or by making a bean trench (see January, page 34).
- Bide your time: don't expose runner beans to night-time temperatures until early June.
- Sow at least two varieties – one white-flowered and one red, because white-flowered varieties are more heat-tolerant than red ones.
- Find a sheltered position out of the wind. If your garden is windy, plant wigwams, not rows. They resist the force of the wind much better. Water the beans well until they get to the top of the canes.
- Once they begin to flower, water them if the weather is dry.
- Once they reach the top of the pole, pinch out the shoots so that they bush out lower down.
- Sow a second lot in June for an autumn crop.
- Pick your crop regularly. Often August is a peak month, so if you plan to go on holiday then recruit a neighbour to pick them and eat them for you.

VARIETIES

'Polestar'
Probably the finest red-flowered stringless runner-bean variety, producing a long succession of thick, fleshy pods until late in the season. Lots of fleshy, tender, smooth-skinned beans.

'Red Rum' AGM
The earliest red-flowered runner bean to crop, producing medium-length, stringless, straight pods. This variety is often over by August, so do grow a late variety too.

'White Emergo'
A white-flowered runner bean producing very smooth, light-green pods.

'Moonlight'
The first runner x French hybrid. Vigorous and long-cropping, producing pods that resemble the runner bean in shape, but when snapped the pods have the plumper profile of a French bean.

'Lady Di'
Red-flowered, stringless runner bean with long, straight pods.

'Firestorm'
A red-flowered hybrid bean that produces plump, long pods.

Organic Tip ✔

Only sow in the middle of May if you feel we are poised on the edge of summer. If mid-May is cold, wait for the weather.

5 Re-sow Early Crops
(late May)

MANY OF the crops you sowed a few weeks ago will be racing away, but if you have room, sow another batch now. They will crop in the autumn and extend your productive season. You will need to water these new rows, as we should be entering the driest, hottest time of the year.

Re-sow three types of lettuce but make sure that one is a Cos: these are slower to bolt than the others. Bolting is normally a lot less likely after Midsummer's Day (21 June), but exceptionally dry summers can still send every lettuce into seed. Your sowings of beetroot can also be repeated, using the same varieties.

Your earliest sowing of carrots included fast-maturing varieties like 'Amsterdam Forcing 3' and 'Nantes Early 2'. Now it's time to change to slower-maturing varieties capable of staying in the ground into autumn and beyond. The following three AGM varieties are all F1 hybrids: 'Bangor', 'Eskimo' and 'Kingston'. Once mature, they will all stand low temperatures. However, if you continue sowing up until mid-August, it's best to switch back to faster-maturing varieties again as these crop more quickly.

Peas can be succession-sown every 2 weeks. The head gardeners managing large walled gardens devoted more room to them than to any other crop for that reason. Peas can be sown up until the middle of July and still crop. In fact, these cool-season crops do well from July sowings. Alongside the varieties listed overleaf, 'Hurst Greenshaft' does well sown late.

SECRETS OF SUCCESS

- Sowing in May provides autumn crops, but watering is more important for these sowings because hot, dry weather often arrives in June just as they are starting into growth. Morning watering is the most efficient.
- If you're sowing any crop in late July, opt for a fast-maturing variety, particularly for root crops.
- Peas enjoy cool weather and can tolerate some shade.
- Pinch out growing tips on later-sown peas and beans as soon as the first pods are ready at the bottom of the plants.

Did you know? Early pea varieties were not very sweet until the Herefordshire cider-maker Thomas Knight started to hybridize peas in the nineteenth century. He used his experience as a cider-apple-grower, selecting and crossing for sweetness, and created wrinkle-seeded peas that could be eaten raw. From 1811 to 1838 Knight was one of the first presidents of what eventually became the Royal Horticultural Society.

VARIETIES FOR SUCCESSION SOWING

Carrot
'Eskimo' AGM
The most cold-tolerant variety, it can be overwintered on well-drained soils. A very strong variety with robust tops. Crown normally below soil level.

Pea
'Dorian' AGM
The long, broad, straight pods contain eight to ten large, tasty peas and this variety was once known as 'Mr Big'. A good crop of pods for picking over a long period.

Pea
'Balmoral' AGM
The ideal pea for a May–June sowing with an autumn harvest in mind. Seven sweet peas in each pod. Dark, handsome foliage and a good yield.

Pea
'Kelvedon Wonder' AGM
This 1925 stalwart variety is still considered the best for regular successional sowings. Heavy crops of narrow, pointed pods in pairs, averaging seven or eight succulent peas per pod. Excellent from a June sowing.

Organic Tip ✔

Be adventurous: push back the boundaries and experiment with late-sown crops because autumns are tending to be warmer and more frost-free than they were.

6 Sow Witloof Chicory

(late May)

THERE ARE lots of different varieties of chicory, including red-leaved and curly forms called radicchio. Witloof chicory is a special type suitable for forcing as a winter vegetable. You must choose Witloof if you want to produce the pale, pointed chicons (as the blanched chicory heads are called). These resemble small Cos lettuces and can be eaten in the darkest days of winter when many other crops may be frozen into the ground. Eaten fresh and shredded into a salad, Witloof has a bittersweet, crunchy texture. When cooked, its bitter flavour balances creamy chicken dishes.

When the crop is in the ground, Witloof chicory looks like oversized dandelions and the tops stay green during winter. It can be sown now, ready to be planted outside in late June or early July. You can sow it outside in a row and then thin, but the plants have to be spaced 30cm (12in) apart, so raising them in pots is easier. Sow seeds in a seed tray, then prick them out individually in small plastic pots. Twenty-five plants would probably be enough for most gardeners. The plants are lifted from November onwards for forcing (see page 334).

VARIETIES

'Witloof Zoom' F1
Go for this F1 hybrid every time because it has the vigour to produce a plump, pale chicon in 4 weeks.

'Variegata del Castelfranco'
This is the hardiest chicory of all. It can be grown and eaten in leaf, or left to heart up, or it can be forced.

'Treviso Rosso'
A red-leaved chicory that can be eaten as a leaf during winter. It can also be lifted and forced.

SECRETS OF SUCCESS

- Seeds can be sown straight into the open ground and then thinned out so that each rosette is 30cm (12in) apart.
- Sowing in pots is easier, but these tap-rooted plants must be transplanted young.
- Keep the plants weed-free and water well in dry conditions.
- You can either select and lift plants as and when you need them from early November onwards, or you can lift all the roots at once, storing them horizontally in peat or sand in a box in a cool shed or garage.
- Force a few of the roots at a time. Up to five roots will fit into a 22cm (9in) flowerpot. This is a convenient size to cover with a bucket. If you use an upturned flowerpot, cover the holes to exclude any light.
- Choose a dark place that's not too warm for forcing – 16–18°C (61–64°F) is ideal.
- Keep the pot of compost or soil moist and warm. Your chicons will be ready to harvest in about 3–4 weeks. Lift a little sooner if you want smaller ones.

Did you know? Witloof Chicory was discovered in 1830 by accident by a Belgian farmer who was growing chicory roots. These were used as a cheap coffee substitute. While storing roots in his cellar, he noticed the new white leaves, tasted them and found them moist, crunchy and slightly bitter.

Organic Tip ✔

Dark conditions are essential for forcing chicory and the flavour is better. Too much light causes the leaves to become over-bitter.

SUMMER TASKS

FRUIT

- Pick your fruit regularly. Begin to train and tie in new shoots of blackberries and loganberries.
- Remove strawberry runners to keep the vigour in the original plants. You could pot up some of the runners should you wish.
- Protect your soft fruit from birds, using netting.
- Cut off strawberry foliage after the plants are finished fruiting. Remove the straw, weed the plot and feed your plants.
- Summer-prune apples and pears in late July or August. This consists of shortening the side shoots.
- Late summer is the perfect time to lightly prune fan-trained, cordons and espalier apples, acid cherries and pears.
- Lightly summer-prune peaches, nectarines and apricots.

VEGETABLE

To Do

Thin crops as needed
Concentrate on watering in the first half of summer
Feed tomatoes, peppers, chillies and aubergines with potash-rich food every 2 weeks
Weed assiduously – a hoe is the most useful tool now
Cloche young cucurbits at night in June
Stop picking asparagus
Cut back herbs to encourage fresh growth
Apply mulch in dry summers
Be vigilant about Cabbage White butterflies and asparagus beetle
Look out for potato blight. Cut away and bin infected leaves
Harvest regularly
Loosen onions and shallots, then lift, dry and store
Check tomatoes – the yield of fruit is at its highest in August

Sow Outdoors

Continue to repeat-sow all outdoor hardy crops
Runner and French beans
Florence fennel
Swiss chard
Japanese onions (August)
Short-term green manures (August)

Sow under Glass

Summer and autumn cauliflowers
Chicory for forcing
Cabbages – red, summer, autumn and winter varieties (June–July)
Spring cabbages (August)

Plant

Tomatoes – outdoor and indoor
Peppers
Aubergines
All cucurbits – cucumber, squash, pumpkin, courgette, etc.
Sweetcorn – in a block
Leeks
Winter brassicas – Brussels sprouts, kale, cabbage, purple sprouting broccoli

JUNE

June can be the softest month of all and those long summer evenings should give you a chance to relax now that much of the hard work is done. Keep on hoeing, weeding and watering, taking care to water well, for if you neglect that in June your yields will suffer, even if it pours for weeks in July. Most crops are raring to go, right up until the longest day, but after that they slow down.

Although June is the start of the summer season, there's often a hungry gap in the vegetable garden because lots of crops sown or planted in spring won't deliver until the end of this month at the earliest. Lettuces and spinach can be very useful at this time of year. Don't despair, though, soon you'll be harvesting baby potatoes and scrumping peas and wondering how you are going to manage to get round to harvesting it all.

When it comes to fruit you'll probably devour your first home-grown strawberries this month and these will taste entirely different from the hard, commercial varieties picked when under-ripe. You may already have eaten your first acid-green gooseberry fool and you can see the apples, pears and plums filling out before your very eyes.

Strings of currants begin to glisten now. Redcurrants are the most translucent of all, epitomizing high summer. If you're new to fruit-growing, arm yourself with a good jam-making pan! Use redcurrant juice as a pectin-rich ingredient for setting strawberry jam. The juice gives the jam a wonderfully rich, red glow. Or make redcurrant jelly, an aromatic addition to lots of meat dishes.

FRUIT

1 Summer-prune Gooseberries

(early June)

THE AIM in fruit-growing is to produce flowering buds and generally this is achieved by shortening the long laterals in winter to promote bushy growth. By summer these bushy growths are long, so now you need to shorten the side shoots on fruit trees and bushes to concentrate their energy into producing flowering buds on short spurs. This is important with gooseberries for many reasons. Firstly it makes the bush less congested, allowing more air to penetrate, and this helps to prevent mildew. Secondly, your fruit will be bigger and better because there is less wood. Thirdly, gooseberries, which are one of the most fecund fruits, can over-crop and exhaust themselves and this may leave them vulnerable to disease.

Summer-pruning consists of reducing the young side shoots back to five leaves: these are easy to see now because this year's wood is light brown. This approach suits cordons, fans, bushes and standards.

Cordons and fans need to be kept in strict shape, so you must remove any deviant wood that does not conform to the trained shape.

Bushes should also be thinned, with all low branches cut away so that they develop a short trunk – about 22cm (9in) in height.

Standard gooseberries must be pruned so that they are kept to a minimal shape. If you allow the top to get too bushy or too large it can snap at the graft in a heavy summer gale.

Did you know? The first gooseberry bushes written about were imported into England from France in 1275 and were planted in Edward I's garden at the Tower of London. Henry VIII was also a fan and imported bushes in 1509. Royal patronage meant that by the end of the seventeenth century gooseberries were being widely grown. Many varieties were bred in the late eighteenth and early nineteenth centuries by British nurserymen and one, 'Hero of the Nile', was named to commemorate Nelson's victory at the Battle of the Nile in 1798.

Organic Tip ✔

Gooseberry sawfly grubs can strip bushes in early summer. However, on many occasions friendly chaffinches and blue tits gather the grubs up for their young. Avoid spraying for this reason.

SECRETS OF SUCCESS

- Pruning now promotes growth because the sap is racing away and the light and warmth are at their best. Very vigorous varieties – for example, 'Winham's Industry' and 'Howard's Lancer' – can over-respond and for this reason should be pruned only lightly.
- With old, congested bushes of all varieties, cut out some old, gnarled branches to open up the shape.
- Shortening the lateral back to five leaves often solves aphid problems, as they tend to feed on the new growth.

VARIETIES

For varieties of gooseberry, see January, page 13.

2 Prune Grapes

(early June)

GRAPE-GROWING is tricky in many of the parts of the UK because we are pushing them to the edge of their range and beyond. However, the warm slopes of the Sussex Downs, areas of Surrey and Kent, and the sheltered Leadon Valley in Gloucestershire are places where vines do grow well in the open. In fact, now that autumns are lasting longer, there are 400 vineyards in England, producing 2 million bottles of wine a year. If you want to dabble in wine-making it may be worth growing outdoor grapes, provided you have a sunny slope and lighter soil. Most of us, sadly, do not have enough room for our own vineyard, but you could aspire to grow one vine for edible grapes.

Grapes need pruning and training or they will produce all leaf and very little fruit. The simplest training method is the rod-and-spur system. A strong vertical is trained and side shoots develop close to the main stem. These are then trained and tied along strong wires. This is the system most often used under glass and against walls outside. Vines grown in open ground are trained using the double Guyot system. With either system, all the serious pruning of grapes takes place in winter when the sap is not running (see page 324). Now is the time to prune the flush of spring extension growth and to restrict the amount of fruit. Vines under 3 years old should not be allowed to produce fruit. Remove any that you see.

In the rod-and-spur system, the main stem – the rod – is pinched out when it reaches the desired height – usually 3m (10ft) or less. In spring the two fat buds to which each lateral – or spur – was cut back in winter grow away strongly. The strongest – the 'heir' – is allowed to grow and fruit; the weakest – the 'spare' – is cut back to a couple of leaves. Now, in June, summer-pruning consists of shortening the small leafy growths. Those without flower are pruned back to five or six leaves. Those with flower trusses are cut back to two leaves after the truss. Allow just one cluster to develop per lateral shoot for the best dessert grapes. Tie in the pruned growth loosely to allow for regrowth. You may need to prune the regrowth later.

Bunches may need to be thinned to prevent overcrowding; use a pair of long, thin scissors. Remove all developing grapes that face inwards and thin outward-facing grapes so that those left can plump up nicely for eating.

For open-ground vines, pruned according to the double Guyot system, the three central replacement shoots are tied in vertically but not pinched out. The fruit-bearing laterals are also trained vertically but are pinched out three leaves above the top wire.

Did you know? Vines have an ancient history. We know that wine was made by the Egyptians from vines growing in the Nile Delta almost 5,000 years ago. Paintings in the tomb of the astronomer and scribe Nakht (*c.*1400 BC) depict vine-growing, with bunches of grapes being trodden by men steadying themselves with ropes.

Organic Tip ✔

Even if you fail to grow edible grapes successfully, a vine on a pergola is a great addition to any garden. The dappled shade cast by the large leaves is perfect for sitting under, and the leaves colour up wonderfully in autumn. The blackbirds are very glad of any small fruit produced.

SECRETS OF SUCCESS WITH GRAPES

- A sunny position and good drainage are vital, so a south-facing slope is the ideal. Cool greenhouses and sheltered walls can also be used.
- Take time to prune and train your vine.
- Restrict the fruit and thin the trusses with grape scissors.

VARIETIES

'Black Hamburgh' (now known as 'Schiava Grosso')
This needs an unheated greenhouse to produce its reliable crops of flavourful black grapes. Widely grown and good to eat.

'Buckland Sweetwater'
An easy, pale-green grape that produces early trusses in abundance. Best under glass and not too vigorous, so suitable for a small greenhouse. May need extra feeding.

'Interlaken'
For a wall or fence. Medium-sized, golden fruit. Thin the bunches for the optimum crop. Good in a cool climate.

'King's Ruby'
Large bunches of dark-purple–mauve grapes with a rich flavour. For the greenhouse or a warm wall.

3 Peg Down Strawberry Runners

(mid-June)

SOME STRAWBERRY varieties produce runners and these will be coming thick and fast now. You should allow only strong, established plants to form runners. If the mother plant is small or very young, cut them away. Plants often bear more than one runner on each stem, but it is best to trim them back to just one runner to conserve the mother plant's energy if you wish to keep that plant for next year. If you are scrapping that plant, you can allow it to run itself out to exhaustion.

Runners can either be potted up into small pots (whilst still attached to the mother plant) or pegged down. The easier of the two is to peg them down into the soil, because then they are largely self-sufficient and need less care. If the weather is hot, small potfuls can shrivel and dry. Lengths of bendy, green-coated wire, readily available from garden centres, are the best material for pegging down, or you can just weigh the runner down with a stone.

Once the runners have rooted well, cut them off from the mother plant and either pot them up or plant them in a new bed straight away.

Keep the young plants well watered. Always resist the urge to let spare runners root into your strawberry bed among your established plants. This will result in a poorer crop.

Keep a look out for vine weevil: the telltale signs are semicircular notches on the edges of the leaves. If you see this, give all your plants a good tug. Any badly affected ones will come away in your hand. Bin them. Use vine weevil nematodes immediately.

Did you know? Strawberries have been a popular food for centuries, but at first they were the wild strawberries from *Fragaria vesca*, which produces tiny red fruits in woodland and hedge bottoms. People used to dig up the roots and plant them in their own gardens. They were used medicinally as a treatment for fever, rheumatism and gout, and the leaves were used as a tea substitute. They are a good source of vitamin C. Fruit was rubbed on to the skin to lighten freckles and soothe sunburn, and they were also used to whiten the teeth.

Organic Tip ✔

Strawberries love rich, friable soil, so double digging a strawberry patch when planting new runners will double your crop. Keep the new bed well weeded and fed with potash (liquid tomato food) between early May and late August.

SECRETS OF SUCCESS

- If you want to produce lots of runners to bulk up a favourite variety, cut off all the blossom as it appears.
- Strawberries are fairly shallow-rooted, so young plants and runners need to be kept well watered until established.
- Eradicate any weeds from your bed because strawberries resent competition.

VARIETIES

For varieties of strawberry, see February, page 42, and March, page 72.

4 Train New Blackberry Shoots

(mid-June)

BLACKBERRIES are the perfect partner for cooking apples and they freeze well too, so if you have room, plant one. Thorniness differs from variety to variety, as does stature – see Varieties, page 44. Some can be accommodated in a container while others will fill an entire corner of the garden, though it's quite possible to train them up a fence, over a trellis, or up a pergola. Thornless varieties like 'Oregon' have decorative cut foliage and are suitable for the ornamental flower garden.

Blackberry flowers are extremely attractive to bees and butterflies – they do not need a partner close by to set fruit. However, fruit set is heavier if your blackberry is in a sheltered position.

The more vigorous blackberry varieties send out new runners during the summer and these will replace the older runners that are bearing fruit this year. The new runners can be treated in various ways. Some gardeners tie them into a bundle in autumn, then untie and arrange them in early spring; others train them in now on stout wires – see page 42. In both cases the old runners are cut out at the base once they have finished fruiting. This encourages vigour.

Did you know? The British blackberry is not a single species but a complex group of 350 micro-species. Some areas have micro-species that have large fruit; others have mean little berries. If you find a good bush, make a note of where it is.

Organic Tip ✔

Blackberries propagate by tip-rooting. If you want a new blackberry, just bury the tip of one branch into the soil. It will root with alarming speed.

SECRETS OF SUCCESS WITH PRUNING BLACKBERRIES

- Equip yourself with strong leather gloves, goggles and long sleeves before doing battle with a thorny blackberry.
- Use sharp loppers and reduce the canes in stages – it's safer (see page 44).
- Pruning encourages vigour, so make it a yearly event.

VARIETIES

For varieties of blackberry, see February, page 44.

5 Look After Your Citrus Fruit

(late June)

ORANGES were once the height of fashion for rich landowners – so much so that special orangeries were built to accommodate them. In winter the oranges were kept warm inside, but in summer the trees were taken outside because the heat of the orangery became too intense for them. This is a good system to adopt even if you do not have your own orangery.

Keeping citrus in good condition needs a careful hand and constant vigilance. They like a gentle regime free from extremes and are not plants for the weekend or casual gardener. At this time of year your orange or lemon tree should go outside. Find a warm, sheltered place away from scorching sun and stand your pot on a tray of wet pebbles. This will help to promote humid air and prevent die-back and leaf drop. Do not allow your tree to dry out. In hot weather, a daily water is advisable and rainwater can be sprayed on the leaves as well. Their love of humidity makes citrus poor houseplants for centrally heated homes, although they can do well in heated conservatories with ventilation.

Fruit ripens over winter, but all leafy growth takes place in summer, so a nitrogen-rich citrus food should be used; special winter and summer formulas are available. Most citrus fruit is trained as a mop-headed tree and summer-pruning consists of pinching back any branches that seem to be growing too strongly in order to maintain this rounded shape. If you need to remove a branch, February or March are the best months.

Did you know? Satsumas are the hardiest citrus of all, able to survive -11°C (12.2°F) for short periods. The variety 'Owari' proved to be the hardiest citrus in a recent trial, surviving several nights when temperatures fell to -6°C (21.2°F) and below.

SECRETS OF SUCCESS

- A young tree will grow much quicker without fruit for the first year, but after that clusters of fruit should be thinned so that each 1m (3ft) tree bears up to twenty fruits.
- Lemons are easier to grow than oranges, but both are tender plants that must be protected from frosts. Night-time temperatures must be at least 11°C (52°F): this will allow the lemon tree to flower and produce fruit throughout the year. This needs to be maintained through winter if possible. Ideally, day temperatures should be 5°C (10°F) higher than this in order to encourage the fruit to ripen.
- Re-pot in late spring if needed, either into the same pot or a slightly larger one. Remove the top couple of inches and tamp in new compost. Use John Innes No. 2 for smaller plants and No. 3 for larger specimens. Add horticultural grit or perlite for extra drainage.
- When watering, allow some water to run out of the pot; this helps to prevent the build-up of harmful salts in the compost.

Organic Tip ✔

Even if your citrus is growing happily in a greenhouse or conservatory, it is still a good idea to move it outside for the summer, because predators (like ladybirds) will be able to frisk the foliage.

VARIETIES

'Improved Meyer'
This is the easiest lemon. The original 'Meyer' was named after the American botanist who brought it back from China in 1902, but it was found to be a symptomless carrier of citrus tristeza virus, a serious disease. 'Improved Meyer' is the virus-free form and should be the one you grow. It reaches 120cm (4ft) and fruits prolifically through most of the year. A cross between a lemon and a mandarin, so not as acidic as many lemons. Foliage is very large.

'La Valette'
A cross between a lime and a lemon, but the fruit is the same shape and size as lemon. Crops well. Compact and easy.

'Quatre Saisons' (syn. 'Gary's Eureka', or just 'Eureka')
Fragrant citrus flowers and fruit through all four seasons. A true lemon that produces lots of fruit. A favourite with many. Can grow large.

'Imperial'
Thought to be a grapefruit x lemon hybrid, it bears much larger fruit than other lemon trees, but fewer of them. Vigorous.

VEGETABLE

1 Watering
(early June)

WATERING the right way at the right time is an art. It's one of the keys to preventing certain crops from bolting, but lack of water is only part of the reason for vegetables running to seed. A lot of crops get stressed by cold spring winds and cold night-time temperatures – something over which we have little influence, although fleecing can help. The best way round these last two problems is to plant later. Every region of Britain has a slightly different optimum time and it can vary within a few miles.

Watering is something most inexperienced gardeners get wrong indoors and out. Seedlings hate being wet and cold, so if you are going to water them, try to get it over with by 4 p.m. Then they have chance to recover and dry off before both enemies arrive on the scenes – the slug and cold night-time temperatures. Invest in a good watering can with a fine rose. Point the rose up for seedlings (to emulate fine rain) and point it downwards to drench larger plants. If seedlings and young plants get too wet, they won't develop a good root system because the roots won't have to search out water.

Don't dribble the hose or can over your garden crops – that only makes the roots come to the surface and they should be heading downwards. Drench your plot for at least 3 hours with a sprinkler or a seep hose. This can be done overnight on warm nights. Three to four bouts of watering are ideal, but even one long watering session will deliver. Always use water wisely and only water if it's vital.

Did you know? The Aztecs used a plant they called *acocotli* (literally 'hollow pipe') to water their crops. This tall plant grew to 10m (30ft) tall and the stems were felled to make pipes to take water to the crops. In 1789 specimens of *acocotli* were collected and sent to the Royal Gardens in Madrid, where the curator, l'Abbé Cavanilles, named the flowers after his Swedish horticultural assistant, Andreas Dahl. The dahlia had arrived in Europe.

SECRETS OF SUCCESS

- Set up an irrigation system in early spring if you have well-drained soil. Use a timer if necessary. Remove it in autumn.
- Seep hoses deliver a slow drip from their sides. They work well as long as there are no kinks or sharp bends in them.
- Porous hoses can be snaked around your plot, as they exude water from all round their circumference. They can be buried to a depth of 10cm (4in).
- Sprinklers are adaptable because you can move them, but avoid using them for certain crops. Some (such as potatoes) are prone to disease if their foliage gets wet.
- Have water butts, as many as you can place, but use this water only on larger plants. It can encourage damping off (a fungal disease) if used on seedlings.
- Tapwater is best for seedlings because it is more or less sterile. It should stand in a can for several hours to warm up. This will also allow some of the chlorine from mains water to escape. Every time you empty a can, fill it up again.

Organic Tip ✔

Mulching helps enormously with water conservation. Grass clippings can be used round large plants like globe artichokes. However, they need to be partially rotted. Lay them on a sheet in the sun until they brown. Turn them once and then apply as a mulch – always on damp soil.

GUIDE TO WATERING

Watering near dusk encourages slugs.

Onions, garlic and shallots have short, stumpy roots that cannot search for water. If these crops become dry early in the year they don't develop.

Beetroot needs water from late spring into summer.

Newly planted leeks are puddled in (see page 52) and this helps initially. One good watering after that helps enormously.

All the cucurbits (squashes, courgettes, cucumbers and pumpkins) love warm, moist conditions and, if the weather fails them, you will need to provide water.

Carrots and parsnips are less fussy – their tap roots help them out.

Potatoes can tolerate drought, although the yield will be smaller.

2 Plant Outdoor Tomatoes

(early June)

THE BEST flavour of all comes from the outdoor-grown tomato, but the British climate is not always warm enough to encourage lots of fruit on traditional tomato varieties. These are often warm-weather plants more suited to being grown under glass. However, in recent years the seed industry has tapped into cold-tolerant varieties from Russia and eastern Europe. These can perform outdoors, although an outdoor tomato is likely to produce around half the fruit you could expect from an indoor one. Generally a crop of four trusses is a good result.

The main problem with growing outdoor tomatoes is their susceptibility to potato blight (*Phytophthora infestans*). This generally appears in August, just when outdoor tomatoes are at their peak, and it usually stops them in their tracks – they just wither and die. Luckily, some of the eastern European varieties are proving very blight-resistant and it is these that gardeners should be seeking out.

Outdoor varieties tend to be bushy and are often best left to develop naturally. The tomatoes are often smaller, but tasty. Plants can be grown in growbags or in pots. I prefer pots: they are easier to move. Plants in south-facing positions against buildings sometimes need moving in summer heatwaves. If you have a very warm position, plant them in the ground.

There are also hanging-basket varieties, but I would grow these in pots too. They are easier to water and they are portable.

Did you know? Until the end of the nineteenth century the tomato was regarded as highly suspicious and possibly poisonous, probably because the flowers resembled those of the notoriously poisonous deadly nightshade, However, cheap glass, which became available in the late 1860s, fuelled their popularity and huge greenhouses were constructed in the Lea Valley in London and along the south coast. A quarter of the tomatoes we eat today are British-grown. Many come from the Isle of Wight – arguably the sunniest place in Britain.

SECRETS OF SUCCESS

- Choose blight-resistant varieties if possible – those with flavour.
- Offer your tomatoes sun and shelter, but don't plant them outside until June.
- Strong plants throw off infection. Water regularly and feed as soon as the first fruit is set, using a liquid tomato food or comfrey tea (see page 154) every 2 weeks.
- Fleece in September if a frost is forecast.

BLIGHT-RESISTANT OUTDOOR VARIETIES

'Stupice'
This cordon variety will need the side shoots pinching out. An early-ripening, cold-tolerant, 'potato-leaved' variety from the Czech Republic. Produces clusters of golf-ball-sized red fruit with a good flavour. A heavy cropper that's ideal for growing outdoors.

'Legend' AGM
This American-bred bush variety has shown great blight resistance in the garden. Produces a heavy crop of large, glossy red fruits.

'Koralik'
This heritage Russian bush variety crops before the main August wave of blight, although it has shown good tolerance to blight in 3 years of trials. The small, bright-red tomatoes on each truss ripen together.

'Premio' F1
This extensively trialled cordon variety produces handsome red fruit with shiny red skin. It ripens well and evenly, and the taste is outstanding.

'Losetto'
A cascading bush variety with exceptional blight resistance that's ideal for hanging baskets, containers or a sunny spot in the garden. Cherry-sized, sweet-tasting, round red fruits.

'Fantasio' F1
A prolific cropper bearing medium-sized red fruit. Vigorous, with healthy green foliage – leave it to its own devices. Grow as a cordon.

Organic Tip ✓

Try to separate your tomatoes from your potato crop to help protect them from cross-infection if blight strikes, as blight spores are carried on the air and by rain splash. This disease also affects aubergines.

3 Sow Squashes and Pumpkins Outdoors

(*mid-June*)

AT THIS time of year it's possible to sow courgette, squash and pumpkin seeds straight into the sun-warmed ground, into a large pot or a growbag. This later planting will soon catch up with the transplanted spring-sown cucurbits and, having been sown now that the nights are warmer, your plants will race away.

The most useful crop of the three is the winter squash. The fruits are cut, ripened outside and then stored for several weeks before being eaten. This resting period in cooler weather allows the starch to turn to sugar. If you are still growing the tasteless marrow, replace it with winter squash immediately!

Squashes need space: they creep over the ground and for that reason they make a good follow-up act after first early potatoes. As they expand they cover the gaps where you've harvested. Potatoes are hungry feeders and their thick foliage tends to exclude rain, so you must enrich the ground before planting squash. Water it thoroughly or wait for rain. Sprinkle on blood, fish and bone (see page 317) and then make a mound 30cm (1ft) high and 60cm (2ft) wide. Space two squash seeds along the mound and then let them fight it out. Pumpkins can be treated in the same way. Alternatively, you can plant some seeds or add a couple of young plants into the top of a well-rotted compost heap – the warmth beneath will give them a boost. If space is tight, sow one or two courgettes instead.

Did you know? One of the most historic squashes is the enormous, teardrop-shaped 'Blue Hubbard'. It was grown in the Americas by native tribes and then by settlers and probably arrived in the USA in the 1700s aboard sailing ships coming from the West Indies. In 1842 a woman named Elizabeth Hubbard mentioned the squash to her neighbour, a seed-trader named James Gregory, who then introduced it as the 'Hubbard' squash.

SECRETS OF SUCCESS

- Warmth, water and fertile soil are essential requirements.
- The large leaves of cucurbits tend to cover the ground and keep moisture in, although you may want to mulch early on.
- Only harvest ripe fruits – the fruit stems should feel woody and corky.
- When cutting, leave a length of stem, then store the fruits upside down on a slatted seat or bench so that they harden and ripen in the sun.
- Grey-skinned squashes store until late April, so use the orange and butternut squashes first.
- Cook squashes from mid-November onwards. Cut them up, then remove and discard the pips. Dice the flesh into 5cm (2in) cubes, adding herbs and oil before roasting.

VARIETIES

'Crown Prince'
A silver-skinned squash with bright-orange flesh. About 5kg (11lb). Stores until March. More useful than 'Blue Hubbard', which is too large for most people – it could feed twenty.

'Potimarron'
An orange, teardrop squash weighing up to 4kg (8lb) that originated in China despite its French name. Good chestnut flavour. Stores until December.

'Hunter' AGM
The butternut squashes are generally difficult to grow in Britain, preferring hotter weather, but 'Hunter' has been specially bred for the British climate. Stores until December.

'Mars' F1
The ideal-sized pumpkin for Halloween carving, with orange flesh that can be used in pies. The green-skinned fruits turn a rich orange as they ripen.

Organic Tip ✔

Always store squashes and pumpkins stem down because moisture can collect in the hollow round the stem, which can lead to rotting. They keep for much longer upside down.

4 Plant Cucumbers
(mid-June)

IF YOU'RE growing lettuce and tomatoes you should aspire to cucumbers too. Outdoor varieties can be trained up and over a compost heap, or fitted into small niches in the vegetable garden – somewhere sheltered that can accommodate a simple 1m (3ft) high trellis. It's still possible to sow F1 seeds (see page 91) but ready-grown plants should also be available now.

Outdoor cucumbers are short and rotund, often with spiny

skins. Normally they measure only about 15cm (6in) in length, but picked young they have a lovely nutty flavour and their compact size means that one can be consumed easily in a sitting. Given plenty of water, it's possible to harvest one every other day in the second half of summer – so they are well worth growing. Cucumbers prefer cool, moist air (ideally with temperatures around 20°C (70°F)) and it's often possible to tuck one plant of an indoor variety at the back of a greenhouse where tomatoes are being grown. Indoor varieties produce long, straight fruits similar to those in supermarkets. Of the two, I prefer the outdoor ones, thinly sliced in wafer-thin white bread on a perfect summer's day.

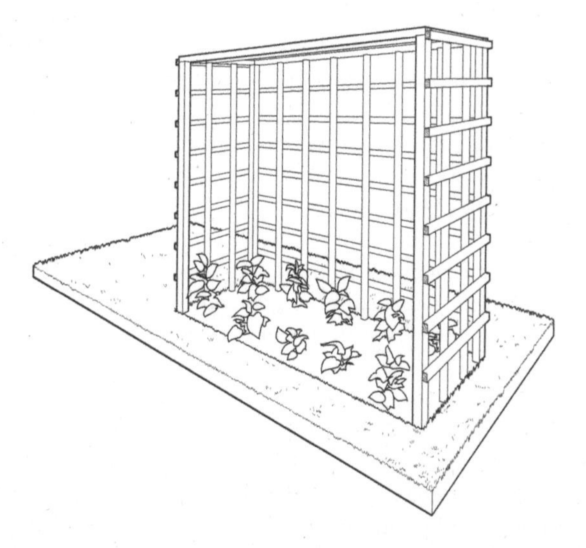

Organic Tip ✔

Don't allow your cucumbers to become old and develop thick, bitter skins. They are more beneficial eaten with their skins on, as all the antioxidants are contained in the dark skin.

SECRETS OF SUCCESS

- All cucumbers are soft-leaved and therefore resent windy conditions. Find them somewhere sheltered, ideally where afternoon sun falls. Midday sun is too much for them.
- Avoid cold nights. Fleece or cover with cloches on cool June nights.
- F1 cucumber seeds are eye-wateringly expensive. When sowing, insert them into compost vertically with the sharp end down to lessen the chance of them rotting in the compost.
- Grow self-fertile all-female varieties under glass. These never produce bitter fruits and you don't have to pick off the male flowers. Never, ever pick off the male flowers on outdoor varieties: they are needed for cross-pollination.
- Sow up until mid-June in small round pots (placing one seed in each) or create a fertile mound and sow into it (see page 186).
- Support with canes and string. Water in well before evening descends, as slugs target cucumbers. Train upwards – away from the ground.
- Feed every 2 weeks with liquid tomato food or home-made comfrey tea (see page 154).
- Water little and often. Stressed cucumbers succumb to mildew, but don't panic if this happens. Ignore it.
- Cucumbers need lots of water, but their stems can rot in wet soil. Sinking a flowerpot into the ground a little distance from the main stem and watering into that works well.

Did you know? The Roman Emperor Tiberius (42 BC–AD 37) had a passion for cucumbers and his gardeners strove to grow them all year long using portable cold frames. Henry VIII felt just the same, and Columbus even took seeds to the New World. The Victorians grew them widely – and the straight cucumber was desirable even then. Long glass jars were used to stop the fruit from curving.

VARIETIES

'Tiffany' F1 AGM
An all-female, vigorous F1 hybrid for the greenhouse, producing lots of dark-skinned 30cm (12in) long fruits. Good powdery mildew resistance.

'Carmen' F1 Hybrid
An all-female, very disease-resistant variety that crops abundantly, producing over fifty fruits in a season. Easy to train and day-length sensitive, which means most flowers are produced when there are 11 hours of day length. Sow after 1 March.

'Masterpiece' AGM
Short, straight outdoor cucumber with tasty, dark-green fruits about 20cm (8in) in length.

'Marketmore' AGM
Stalwart outdoor variety, producing a good yield of dark-green fruits. A consistent, disease-resistant cucumber.

For further varieties, see April, page 131.

5 Stop Harvesting Asparagus

(late June)

FOR NOTES on planting and cutting asparagus, see page 296. The asparagus season ends on Midsummer's Day or thereabouts, and cutting should now stop to allow the plants to regenerate for next year's crop. This shallow-rooted crop cannot be hoed very easily, so hand-weed the bed and lightly sprinkle blood, fish and bone round each plant. Do not overfeed asparagus, however. Watch out for female plants that produce seeds (these look like small berries) and remove any you see to prevent inferior seedlings popping up. Most modern varieties are all-male, so in theory it shouldn't happen. Stake any over-tall stems to stop them flopping.

Keep an eye out for asparagus beetles: they look a bit like elongated ladybirds, with orange, black and white bodies. Pick any off and destroy them, and check regularly for more. Leave the top growth intact at the moment, but cut it down when the foliage starts to fade to 2.5cm (1in) above the ground. This treatment will give your crop 4 or 5 months to recover for next year.

Asparagus is very long-lived and it can tie up the same plot of ground for decades. You need thirty crowns to stand any chance of a harvestable crop, so it isn't an option for small gardens – particularly when you realize it generally has only a 6-week season. However, if you have a large enough plot and you love it, grow it. It arrives in the hungriest moment of the gardening year, when most crops are getting established, and that's why country-house gardens still grow so much of it.

Did you know? Asparagus produces rather phallic spears and the Greek *spargao* (meaning 'turgid') reflected their tumescent shape. The sandy land close to Venice produced masses of white asparagus in the sixteenth century. When the Huguenots fled from France in 1685 they brought it with them and planted fields of it. Battersea asparagus was rated the best and the earliest, and 260 acres were devoted to the crop close to the river Thames.

SECRETS OF SUCCESS

- Plant asparagus only if you have room for thirty crowns. Otherwise you will never get a decent-sized crop.
- This crop requires great patience. You have to stand back and allow at least 3 years before you start cutting any spears – longer if you are raising plants from seed.
- Asparagus thrives in warm places on well-drained, fertile soil, so river banks with sandy, alluvial soil are perfect. Many gardeners on heavier land compensate by making a raised mound containing coarse grit and organic matter to provide adequate drainage.
- Plant new crowns in April on well-prepared, enriched and weed-free soil. Soak them for an hour or two first (if they look dry), then spread out the roots to form a wagonwheel shape. Space each crown 30cm (1ft) in rows over 1m (3ft) apart to a depth of 10cm (4in).
- Avoid planting in frost pockets.
- Try to pick every other day, but crops depend on warm weather.
- Always feed in March and avoid mulching. Asparagus is shallow-rooted and resents being mulched in spring.
- Mulch after cutting has stopped (in mid-June) and again in autumn. A 5cm (2in) layer is enough.

Organic Tip ✓

If you've suffered from asparagus beetle, don't add the stems to your compost heap – bin them. The newly hatched beetles overwinter in debris and the compost heap is a perfect place to shelter before emerging in the spring. In May and June the grey eggs are laid haphazardly along the stems and can be rubbed off easily with your finger.

VARIETIES

'Connover's Colossal' AGM
Still highly popular despite being launched in 1872. An early, heavy-cropping variety and the green spears do well on light soils.

'Backlim' F1 AGM
A mid- to late-season male variety giving consistent yields of large, smooth spears with well-closed tips. A very reliable, healthy performer.

'Gijnlim' F1 AGM
An early-season, high-yielding male variety producing medium-thick, mid-green spears with closed purple tips.

'Purple Passion'
A mid-season, purple-speared variety with tender, supersweet young spears that can be eaten raw in salads.

6 Plant Aubergines

(late June)

AUBERGINES are a demanding crop to grow in the average British summer, but you can encourage them to produce a reasonable crop in containers placed in a sunny position. Use a rich, loam-based compost – John Innes No. 3 is perfect for vegetables – and find a bucket-sized container. Black plastic (which tends to absorb the heat) is fine.

You can also grow aubergines in the ground under cover, either in a greenhouse or a polytunnel. This system will give you at least twice as many aubergines as a container-grown plant. Aubergines seem to have challenged gardeners for centuries. The Moors used to force them successfully using hotbeds over 1,000 years ago. The heat was provided by decomposing manure – a technique also used to raise pineapples in Britain in the mid-nineteenth century.

You can make a simple hotbed by excavating a deep trench and filling it with organic matter, or animal manure. This will decompose and, as it does so, will warm up the soil. If you can get some overhead cover, this will trap any heat and create a hothouse effect. So excavating a bed in the greenhouse or polytunnel will work well.

However, you can also cover your hotbed with a cold frame. The temperature in the frame is on average 10°C (50°F) hotter than the outdoor temperature, and the difference between daytime and night-time temperatures is less extreme too. With this arrangement your aubergines can go on to the covered hotbed in April and be up to 120cm (4ft) tall by June.

SECRETS OF SUCCESS

- Choose an early F1 variety – the ones that produce big, round fruit are the best. The cylindrical fruits do not cook as well: they turn grey.
- Warmth, water and a sheltered position are the key ingredients for this tomato relative. Try to provide all three.
- Feed with a weekly dose of liquid plant food, either tomato feed or home-made comfrey tea (see page 154).
- Keep well watered. Protect from slugs and snails, which will eat the young fruits. Slug hunts at dusk are a must.
- Pick the fruit regularly, even if it's small.
- Dry air means your aubergines will almost certainly fall victim to spider mite. Spray regularly with water or use a biocontrol.

Organic Tip ✔

Ensure the ground is moist around your plants to keep down red mites. Aphids can also be a problem – but allow your predators to deal with them.

VARIETIES

'Bonica' F1 AGM
Glossy, black fruits early in the season on a tall, vigorous plant. Best in the ground.

'Black Enorma' F1
Huge, egg-shaped, dark fruits. Very prolific and carries on cropping late.

'Fairy Tale' AGM
A new compact variety with egg-shaped fruits in purple and white. Good on a sunny patio.

'Bellezza Nera'
Beautiful, fluted, wide fruits that cook equally beautifully. Very large and dusky purple rather than black. Not the same as 'Black Beauty'.

Did you know? During the Renaissance (1300s–1600s) the Italians thought the aubergine was poisonous and evil. It became known as *mala insana* – 'the unhealthy apple'. In other parts of Europe eating aubergines was suspected to cause madness, leprosy, cancer and bad breath. Louis XIV asked the famous French gardener Jean-Baptiste de La Quintinie (1624–88) to grow them at Versailles and by the eighteenth century the aubergine was an established food in Italy and France. The influential cookery writer Elizabeth David introduced them to the British in the mid-twentieth century.

JULY

The growth spurt that was so obvious in spring and early summer has slowed down and the further north you are, the less time you have to replant and gap up for autumn and winter crops. So every time a gap appears, plug it with a catch crop like lettuce, spinach or dwarf French beans. Although these will take longer to produce a crop now that daylight hours are waning, they should be harvestable by September if you choose the correct variety.

The weather this month will dictate whether you'll be weeding or watering and it's important to do both as required. Fortunately the evenings are still light. Keep harvesting your crops, especially the first beans, to encourage more. Enjoy your glut of home-grown produce, for July is one of the most productive months of all.

This is the month for harvesting soft fruit and, once picked, it's possible to summer-prune; this consists of shortening the side growth, or laterals. After the gooseberries, you can tackle the blackcurrants and redcurrants, but they will have different pruning regimes because blackcurrants fruit on new wood and redcurrants on older wood (see page 229).

Stone fruits (such as peaches, nectarines and almonds) can be lightly pruned if the crop allows. Apricots, however, are less prone to disease and they could be pruned in early spring or late summer. It's a little too early to summer-prune apples and pears, so tackle that in August.

FRUIT

1 Look Out for Woolly Aphids

(early July)

KEEP AN eye out for woolly aphids on your apple trees. These brown-black pests produce a waxy white fluff that almost looks like mould. It can appear from late spring onwards and usually persists until early autumn. At first the aphids live in cracks and crevices on the bark, and sometimes on pruning cuts. They feed on the tree sap to begin with, but then they move to soft new growth. In extreme cases the trees develop soft, lumpy growths that are very visible in winter; these are caused by the chemicals secreted into the plant as the aphids feed. These bumpy lumps tend to split in frost and this is when apple canker can take hold. You can buy canker paint and badly affected trees may be improved by pruning, but this will depend on where the lumps are – with cordons and espaliers they often develop on the trunk.

Woolly aphids are extremely difficult to get rid of because they tuck themselves under bark, or in the crevices. There are biological controls, but I prefer the organic 'hands-on' method. Now is the time to scrub them off with a stiff brush, or you can hose them off. This lessens the numbers and limits the damage. Also, in a healthy, spray-free garden there should be plenty of natural predators waiting to pounce. Lacewings are excellent predators of pests in trees.

Did you know? Woolly aphids overwinter on their host plants as immature nymphs. They hide in cracks in the bark or in crevices around old feeding areas. There is no telltale white fluff in winter.

Organic Tip ✔

Bacterial canker is a serious fungal disease of apples that causes die-back and it mostly starts near pruning scars. Keep an eye out for any dead wood, removing and destroying it as soon as you see it.

SECRETS OF SUCCESS

- Remove any infected soft shoots showing the characteristic white fluff produced by aphids now and burn or destroy them.
- Always use sharp secateurs, loppers and pruning saws. Wash them after use, then sterilize them with boiling water, or bleach them, between trees.
- If removing a limb, cut into the bottom of the trunk before making the cut from above, to give yourself a clean cut. Get a helper if it's a large limb.
- Always try to cut into clean, white wood where possible.
- If you have exposed a large area, apply wound paint.
- If you spot dead wood, prune it out and destroy it. It may be canker and special canker paint can be applied.

2 Summer-prune Pears and Apples

(early July)

PEARS ARE not such reliable croppers as apple trees in the British climate. They need warmth in the weeks following pollination to encourage pollen-tube development, which enables the fruit to form. They also like more moisture than apples. Although we can provide extra water, warmth is in the lap of the gods. However, by July your pear crop should be developing well if the fruit managed to set in the first place. Thin the fruit if the tree is still laden after the natural June drop – see page 146. Sharp scissors are the best weapon. Aim for two fruits per cluster as less thinning is needed for pears than apples.

The new growth will be obvious by July. Summer-pruning of pears and apples consists of shortening the laterals (or side shoots of the main branches) that you winter-pruned (see pages 13 and 15). New lateral shoots developed this year are cut back to two or three leaves, although you will have to work round the crop.

Did you know? Pears like heavy soil and the best pear-growing areas are often close to rivers. Waterperry near Oxford, obviously a pear-growing area for centuries, lies low down by the river where the air is kept mild by the flowing water, which acts as a storage heater. When wine was in short supply in the Napoleonic Wars, sparkling perry became the preferred tipple.

Organic Tip ✔

Pears produce their blossom early and spring frosts often prevent a crop. If you have a warm slope, plant your pear there. The frost should slide down to the bottom of the slope and miss your blossom. Like apples, pears are divided into four pollination groups and if you have a cold garden you should plant those in Pollination Group D, sometimes called the late group.

SECRETS OF SUCCESS

- If your garden is cool, opt for 'Conference' – a self-fertile pear with elongated fruit that wants to crop heavily.
- Some pears are self-fertile but many need a partner in the same pollination group. Pears are not as widely grown as apples, so you will almost certainly have to plant two varieties yourself rather than relying on pollen from neighbouring trees.

VARIETIES

For varieties of pear, see January, page 18, and September, page 260. For varieties of apple, see January, page 14, and September, pages 260 and 263.

3 Plant and Summer-prune Kiwifruit

(mid-July)

THIS IS a good time to plant a kiwifruit and also the time to prune an established one. However, I live in the cold heart of the country and I have never managed to raise one edible fruit despite having had a vigorous self-fertile plant that rambled very prettily over a shed for 10 years or more. It shrugged off cold winters – kiwis are actually hardy to −8°C (18°F) – but it never produced any flower or fruit.

If you have a warm, sheltered garden in good, bright light, you are much more likely to succeed. Kiwi plants are mainly dioecious, however: they have either male flowers with feathery, egg-yolk-yellow-tipped anthers, or female flowers with pure-white middles and strong stigma. Some modern varieties are self-fertile and these are popular because you need only one plant.

The key to getting fruit is correct summer-pruning to restrict the rampant growth. The flowers should appear in early summer along the length of one-year-old wood and at the base of new shoots. Once the framework of branches has been established, cut back the fruiting shoots to five or six leaves. This diverts the plant's energies into producing fruit buds and, with less leaf, the developing fruit is exposed to sunlight. Fruit-bearing shoots can be cut to five or six leaves past the kiwifruit. There is no need to thin the fruit.

After the fruit has been harvested, those shoots need to be shortened back to 5–7.5cm (2–3in) in length to encourage more fruit buds.

Did you know? The kiwifruit, *Actinidia deliciosa*, is native to southern China, not New Zealand. Cultivation spread from China in the early twentieth century when seeds were introduced to New Zealand by Isabel Fraser, the principal of Wanganui Girls' College, who had been a visiting missionary in China. The seeds were planted in 1906 by a Wanganui nurseryman, Alexander Allison, and the first New Zealand fruit was harvested in 1910. People who tasted it thought it had a gooseberry flavour, so it was called the Chinese gooseberry. It did not become known as kiwifruit until *c.*1960, when New Zealand growers thought the furry brown fruit resembled their national bird, the kiwi. Italy is now the leading producer, followed by New Zealand, Chile, France, Greece, Japan and the United States.

SECRETS OF SUCCESS

- Buy well-grown plants with a strong main stem and space them 3m (10ft) apart. If they are not self-fertile, plant a male and a female.
- Plant them in good light in an open, sunny position.
- Kiwifruit need well-drained, moisture-retentive soil, so prepare the ground well when planting.
- Summer moisture swells the fruit, so water well in the growing season. Mulching helps once spring warms up, but avoid contact with the main stem.
- Feed with a high-potash feed in late winter, then use a general fertilizer.
- Cold, wet soil damages the roots in winter, so those on clay soil will have great problems.

Organic Tip ✓

This is a fruit for a warm garden! It needs a long growing season in order to ripen well. If you can grow it, and then eat it, it will boost your vitamin C levels enormously.

VARIETIES (NB ONLY FEMALE OR SELF-FERTILE PLANTS BEAR FRUIT)

'Hayward'
The most widely grown female kiwi. It is very late-flowering and produces large, broadly oval fruits with good flavour.

'Tormuri'
This late-flowering male cultivar is suitable for pollinating 'Hayward' (above).

'Jenny'
A self-fertile variety, it can produce well-flavoured fruits.

'Issai'
This hardy kiwi belongs to a different species, *A. arguta*. It bears small fruits about the size of a grape in July or August. These are eaten whole.

4 Order New Apple Trees for Autumn Dispatch

(late July)

THIS IS an excellent time to think about ordering new fruit trees and bushes because specialist nurseries will be producing stock for autumn and winter delivery. At this time of year the choice is greatest and because popular varieties sell out quickly, it's worth making decisions and buying now. Ask the nursery to send them out in the fruit-planting season, not in summer.

Nursery stock is either container-grown or bare-root and often both are available. On most soils the container-grown tree is best planted in early September. The ground is still warm and the days still clement, so a newly planted tree will settle in before next spring. However, if you garden on clay soil my advice would be to wait until spring, because heavy soil is wet and cold in winter. You could also improve the ground by adding organic matter or coarse grit. Another technique, often used by Victorian gardeners, is to plant on a raised mound so that the water has somewhere to drain away.

Bare-root trees are sent out when dormant and can arrive between November and March. If you order now you will be at the top of the list and should get your trees in November. Prepare the ground before planting and cover it up with cardboard or old carpet to keep out the frost, so that when your tree arrives you can plant it out on the first clement day.

For general advice on planting, see page 7.

Did you know? Lots of rootstocks were used by our ancestors, but the naming was very muddled. For example, as many as fourteen different kinds of rootstock were all labelled 'Paradise'. In 1912 Ronald Hatton and Dr R. Wellington of the East Malling Research Centre in Kent began to sort out the incorrect naming. In 1917 the John Innes Centre, then based at Merton near London, joined with East Malling and they began a breeding programme to produce better rootstocks. The Malling–Merton series is still used with apples today – see Types of Apple Rootstock, opposite. 'MM' means that the rootstock came out of the collaboration and a single 'M' denotes that the rootstock was developed at East Malling.

Organic Tip ✔

If you are a serious organic gardener, make disease-resistance your top priority, but do make sure you enjoy eating the fruit too.

TYPES OF APPLE ROOTSTOCK

MM111 – vigorous
Suitable for a wide range of soils and staking not necessary if one-year-old trees are planted. The mature height is 5m (15ft) and the yield at maturity is 45–180kg (100–400lb) of fruit. Fruit appears after 3–4 years. Used on standard, half-standard and large espalier apples.

MM106 – semi-vigorous
Suitable for a wide range of situations and soils – even poor soil. Trees reach an average of 3–4m (10–13ft)) with a spread of up to 4m (13ft). Stake for the first 5 years. Fruit is produced after 3–4 years and the yield at maturity is 23–56kg (50–100lb). Used on half-standard, bush, cordon, espalier and container.

M26 – semi-dwarfing
Can be grown in all reasonable soil conditions, including grassy orchards. Stake for the first 5 years. Trees reach an average of 2.4–3m (8–10ft) when mature, with a spread of 3.6m (12ft). Fruit is produced after 2–3 years and the yield at maturity varies between 13.5–36kg (30–80lb). Planting distance is 2.4–3.6m (8–12ft) apart with 4.5m (15ft) between rows. Used on bush, pyramid, centre leader, cordon, minaret, espalier and container. Suitable for small gardens.

M9 – dwarfing
Needs good fertile soil and weed-free ground. Permanent staking is required. Water in dry conditions. Trees reach an average of 1.8–2.4m (6–8ft) with a spread of 2.7m (9ft). Plant 2.4–3m (8–10ft) apart. The yield at maturity will be 11–23kg (25–50lb). Fruit appears after 2 years. Used on bush, pyramid, centre leader and cordon.

M27 – extremely dwarfing
Requires fertile soil conditions and the ground needs to be weed- and grass-free. Permanent staking required. Water in dry conditions. Trees reach 1.2–1.8m (4–6ft) in height with a spread of 1.5m (5ft). Plant 1.2–1.5m (4–5ft) apart with 1.8m (6ft) between rows. The yield at maturity is 4.5–7kg (10–15lb) of fruit, and fruit appears after 2 years. Used on dwarf pyramid, centre leader and stepover apples, but not on weaker-growing varieties.

For varieties of apple, see January, page 14, and September, pages 260 and 263.

SECRETS OF SUCCESS

- Research your local varieties by telephoning the managers of specialist fruit nurseries, going to 'apple days' and to local gardens.
- Don't be too tied up on heritage fruit varieties. Explore the modern too: breeding in North America is producing cold-tolerant varieties that will do well in the UK.
- Prepare the ground well, then stake and tie in the tree as you plant.
- Rootstocks govern the vigour of trees. Apples vary greatly in habit, so do ask about which to grow – your experienced nurseryman will know which varieties do well on which rootstock.

5 Summer-prune Cherries, Plums, Nectarines and Peaches

(late July)

THESE FRUIT trees are not pruned in winter due to silver leaf disease – see page 45. Instead they get sympathetic pruning now so that the sap seals the wounds.

Plums should first have any diseased or dying wood removed. Then, leaving the main leading shoot at the apex of the tree alone, shorten the branch leaders (the growth at the tips of the branches) by 15cm (6in). The side shoots or laterals are cut to three leaves and any sub-laterals (secondary shoots from the laterals) are cut back to one leaf. This maintains the shape of the tree – usually pyramidal.

Peaches and nectarines need very little pruning because they crop on young wood produced in the previous year. New shoots should be retained every 10cm (4in) along the leaders. Do not prune these leaders. However, cut out any other surplus shoots to roughly 2.5cm (1in) to relieve overcrowding.

Cherries are divided into sweet dessert cherries and acid cherries and both are pruned after the fruit has been harvested. Sweet cherries fruit on one- and two-year-old wood and on spurs of older wood. As a result, they require lighter pruning. Remove any dead, damaged or diseased branches. Then remove very weak and badly placed wood. Shorten the tips of the remaining branches by about a third of their new growth to help encourage the development of fruit buds. Cut out any side shoots that are over 30cm (12in) in length and thin out very crowded shoots. Leave side shoots that are less than 15cm (6in) long unpruned. You can shorten other side shoots to five or six buds to encourage a succession of fruit over the next 2 years.

Acid cherries, which produce dark fruit that is usually cooked rather than eaten raw, are more vigorous and are pruned and thinned more rigorously, as they fruit on the previous year's wood. Remove any dead, damaged or diseased branches and any weak or badly positioned branches that are rubbing together. Then remove about a quarter of the remaining older wood, cutting back to a main branch or younger side shoot. This reduces overcrowding and encourages new growth.

Organic Tip ✔

Wall-trained stone fruit is much easier to net against birds. It's also easier to cover up in winter to prevent disease.

Did you know? The Romans are thought to have introduced the sweet cherry to Britain, and they in turn got it from Turkey in the late first century AD. Archaeologists, though, have found cherry stones in sites all over Europe that date back to 1000 BC and beyond. The wild cherry grows throughout Europe and produces pleasantly edible fruit. What is odd is that some cultivated varieties have an extra chromosome – seventeen instead of the sixteen of the wild cherry. Perhaps the Romans did after all obtain an ancient hybrid from Turkey, which is still the world's largest producer of cherries today.

SECRETS OF SUCCESS WITH STONE FRUIT

- Look for modern varieties resistant to peach leaf curl (see page 116).
- Seek out the dwarfing cherry rootstock Gisela 5. Trees and fans will reach 3m (10ft) in height when mature. At the moment only 'Stella', 'Sunburst' and 'Morello' (an acid cherry) are available. All cherries grown on any Gisela rootstock need permanent staking.
- Acid cherries (including 'Morello') can produce a crop on shady north-facing or east walls.
- North American and Canadian breeding is producing cold-tolerant varieties, but generally most stone fruits love warmth.

VEGETABLE

1 Cover Up Cauliflower Curds

(early July)

CAULIFLOWER is the hardest vegetable of all to grow. It's the most susceptible to club root of all the brassicas, so it should always be raised in modules. It's more attractive to birds, slugs and caterpillars than any other, so it must be netted. It also demands good growing conditions (with lots of nitrogen) and it needs watering and feeding when growing. Finally, it takes up a lot of space. But a home-grown cauliflower is a sweet delicacy and quite different from one that's been commercially grown.

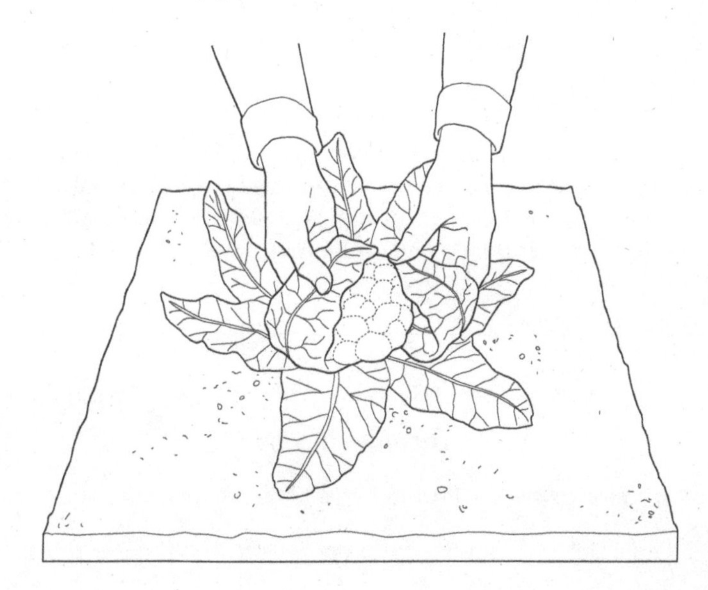

There are three types of variety: summer, autumn and winter. However, most gardeners opt for autumn-cropping varieties that can be harvested in September after the peas and beans have waned. These cauliflowers are sown in April and planted outside in June.

If you're prepared to take the time and trouble, the following technique works. Sow your seeds in modules (one seed in each) and once the plants reach 7.5cm (3in) in height, carefully remove them and put them into small individual pots measuring 7.5–10cm (3–4in) in diameter. Grow these on for a further 2–3 weeks, keeping them well watered.

Prepare the bed well, adding a granular, nitrogen-rich fertilizer just before planting. Tap one or two plants out from their pots and check that the roots have not begun to circle the pot. The roots should be just touching the edges. If they are circling, tease them out. Space each plant 60cm (2ft) apart and net against butterflies.

Carefully nurture the plants, making sure they have enough water. Then, as the curd develops, fold the large outer leaves over the curd to cover it (see previous page). This keeps it white – exposure to light can turn it yellow.

Did you know? The cauliflower is thought to have come from Cyprus in the sixteenth century and was considered as exotic as the melon then. Josiah Wedgwood made a cauliflower teapot in 1760, one of the first novelty teapots ever made.

Organic Tip ✓

Variety is everything and the more compact F1 varieties are superior.

VARIETIES

'Gypsy' F1
Officially a summer variety, but also works well for autumn, and this one crops over 2–3 weeks.

'Pavilion' F1 AGM
A very uniform, Australian-bred variety with a deep, white curd – the favoured choice of many.

'Graffiti' AGM
Amethyst-coloured curds in the second half of summer. Good to eat, and keeps its colour quite well when steamed. Broccoli flavour.

'Romanesco-Veronica' AGM
Easier to grow than most. Whorls and peaks of bright-green curds. Broccoli flavour.

SECRETS OF SUCCESS

- Don't grow too many. Eight plants are enough for most gardeners because cauliflowers tend to crop all at once.
- Don't allow them to be checked by lack of water or food – unless you want a golf-ball-sized curd.
- Pick them when they are perfect as they go over within days, losing flavour and texture.
- Remove the whole plant when harvesting, otherwise it will use up food in the soil as it re-sprouts.
- The white-curded varieties are more difficult to grow than the more colourful varieties.

2 Fill Gaps with Dwarf French Beans

(early July)

AT THIS time of year gaps start appearing in patches rather than whole rows and the wise vegetable gardener should make use of every inch by sowing catch crops that deliver quickly. Best of all for this purpose are dwarf French beans. Certain varieties can crop in 7 weeks and there's no staking. The nitrogen-fixing roots will also add fertility to your soil, and as soon as the crop is over the beans can be removed easily. Just a few plants produce a heavy crop of beans, enough for several meals, and it's quite possible to lift a single root of potatoes and sow a few beans straight away. Dwarf French beans crop for roughly 4 weeks and probably occupy the ground for 12 weeks or so, so they will be gone by the time autumn preparation starts.

French beans enjoy warm sun and crop more heavily in warmer conditions. I enjoy their flavour and crunchy texture when lightly cooked and added to salads. Pod colour varies between purple, yellow and green, making them look summery on the plate and in the garden, although sadly purple beans cook to dark green.

If you are plugging a gap, push a few seeds into the ground to a depth of 5cm (2in). Space them out roughly 10cm (4in) apart. Water them well after sowing. If you have enough room for a row, space two rows 38cm (15in) apart and sow them in a shallow drill, leaving 2.5–5cm (1–2in) between each seed.

Did you know? Although called French beans, these plants originated in South America, where they have been eaten for millennia. They arrived in Europe in the early sixteenth century and, as they became popular in Italy first, they were called Roman beans. However, by the mid-nineteenth century they had become known as French beans, possibly because more breeding was being carried out in France.

SECRETS OF SUCCESS

- Choose a warm, sunny site.
- Slugs adore French beans and the dwarf varieties are so close to the ground that they present little challenge to gastropods. Hoe between plants if possible and always water in the first half of the day – well before slugs become active.
- Pick carefully, because the large leaves hide the beans and if any run to seed it will slow down your crop.
- Green and yellow varieties are speedier to crop than purple-podded ones.
- Water once flowering starts, if it's dry.

VARIETIES

'Stanley'
The fastest-maturing bean I have grown, producing a heavy crop of slender green beans in one flush – which is useful.

'Sonesta' AGM
Compact plants producing a high yield of yellow, waxy beans.

'Purple Tepee'
Purple pods 15cm (6in) in length with a very good flavour. The name is misleading: it does not form a tepee – it's a compact dwarf French bean.

'Delinel' AGM
Slightly mottled green beans on strong plants. Exceptional flavour.

Organic Tip ✔

Only plant dwarf French beans in sunny, warm positions. If you have a shadier gap it won't be warm enough – sow peas or spinach instead.

3 Sow Florence Fennel

(mid-July)

FLORENCE fennel provides a real taste of summer with its aniseed flavour. It's also handsome, with swollen cream roots and green feathery foliage. However, it's a tender annual and therefore can be sown only between May and July. It needs warmth and can fail in poor British summers. It also has a tendency to bolt (run to seed) if stressed, but it is worth the gamble. Cold nights encourage bolting and May can be chilly. I prefer to sow it after the longest day when temperatures are more consistent.

Warmth and water are needed in order for the bulbs to swell to a good size and this normally takes 10 weeks from sowing. Gardeners in very warm places can get bulbs that weigh over 500g (1lb). Prepare a drill and sow the seeds directly into the ground, to a depth of 2.5cm (1in), thinning them out after germination to leave 25–30cm (10–12in) between each seedling. Sowing direct is more successful than raising plants in modules.

Once the seeds have germinated, water regularly in dry weather but don't drench your plants because Florence fennel prefers good drainage. Little and often is the key. Fleece in September on cold nights, for this is a warm-season crop, as the name Florence suggests. These are demanding plants to grow, but they crop quickly and the flavour is intensely aromatic when the roots are baked or braised so that the edges caramelize.

Did you know? Florence fennel has been known in England since the eighteenth century when a London nurseryman included it in his seed list. He imported the seeds from Italy, where it is widely grown, but our weather makes growing it rather riskier.

SECRETS OF SUCCESS

- Water regularly – little and often – because this crop enjoys moisture and good drainage.
- Light soil is an advantage.
- Earth the bulbs up (i.e., mound soil up against the developing bulbs) as they develop to keep the skins pale.
- Cut above ground level and then the base will sprout and provide fennel-flavoured foliage which can be used to flavour chicken and fish dishes.

VARIETIES

'Victoria' F1 AGM
Well-filled green-and-white bulbs.

'Zefa Fino' AGM
Fine, feathery foliage and creamy, medium-sized bulbs. Matures quickly.

'Amigo' F1 AGM
Uniform, medium-sized bulbs that swell up early in the season. A paler variety.

'Finale'
Slow to bolt, producing wide, white bulbs with green, feathery leaves.

Organic Tip ✔

Mulch your seedlings once the bulbs begin to form – partially rotted grass clippings will do (see page 183).

4 Make Late Sowings of Salads and Roots

(mid-July)

THERE IS still enough time to sow carrots, beetroot and salad crops for late-autumn cropping, especially now that winter usually comes later.

By now most early potatoes are out of the ground, or partly out. If sowing or planting after potatoes, rake the area thoroughly and water the ground well if it's dry. Then apply blood, fish and bone (see page 317) to boost the nutrients. Potatoes leave the soil exhausted and their foliage often prevents rain from penetrating the ground.

Lettuces sown straight into the ground should germinate within 10 days unless we have a heatwave. Hot temperatures delay the germination of lettuce seeds, so be prepared to repeat-sow if necessary. You can also sow seeds in 6 × 4 modular trays, one or two in each module, and then plant those out once their roots have reached the bottom. Good varieties for this time of year include Cos lettuces – the narrow, slender ones with an upright habit. These are less likely to bolt and turn bitter than other types and some varieties show excellent mildew tolerance. These lettuces can be eaten from September until November if they are watered during August, so they are well worth the effort.

Beetroot and carrots also grow well in warm, damp conditions and the second half of summer often provides the perfect climate. However, if August is dry you will have to water both. Sown now, they will produce young roots in late October. Certain winter-hardy carrots can be left in the ground and then dug as needed.

Did you know? The earliest lettuces were almost certainly weeds, because gardeners allowed weeds to develop between their crops and then harvested the weeds whilst they waited for the crops. The Cos (or Roman) lettuce is depicted on Egyptian tomb plaques dating back to 3000 BC and it was eaten and used as a medicinal cure-all.

Organic Tip ✔

When harvesting leafy crops like lettuce and cabbage, always lift the whole plant before cutting. A stump left in the ground can harbour diseases.

SECRETS OF SUCCESS

- Vigorous F1 varieties of carrot and beetroot are best for summer sowing because their growth and germination rate are much faster than others'.
- Be extra rigorous about thinning these crops as they need to mature quickly. They do not have time to fight it out.
- Water well.
- Harvest the first roots when still small by taking from all along the row – this will leave more space for the others.
- Earth up the roots as you harvest – this helps in dry weather.

VARIETIES

Beetroot
'Alto' F1 AGM
Cylindrical beetroot that pushes up as it develops, making it easy to see the size of the root. Sweetly flavoured.

Carrot
'Eskimo' F1
Medium-length orange roots with crowns that stay just under the soil. Very good at overwintering in the ground.

Cos Lettuce
'Lobjoit's Green Cos' AGM
This slow-to-mature, large Cos forms an open head, but the flavour never seems to get bitter. Good late into the year.

Cos Lettuce
'Claremont' AGM
A medium-sized Cos lettuce that resists bolting and mildew. Crisp, crunchy and sweet.

5 Sow Spring Cabbages
(late July)

IT'S TIME to sow spring cabbages for harvesting from next March onwards. These can either provide open heads of rich green leaves (as in spring greens) or in the case of some varieties be left to head up to form small conical heads. Spring cabbages fill the gap between Brussels sprouts and purple sprouting broccoli, although this does depend on weather. They provide tasty leaves with a sweet flavour and can be harvested until June, before other crops are ready.

Seeds can be sown in modules or in a seedbed in the ground. However, seeds won't germinate in high temperatures, so if you are using the greenhouse try to find a cool position.

Seedlings should be planted out once they have five or six leaves – usually when 10cm (4in) high. Early September is the optimum moment, although those with warm gardens can get away with planting in the first half of October. Compact varieties can be planted 30cm (12in) apart; larger varieties are best spaced 45cm (18in) apart.

Brassicas prefer firm soil and spring cabbages should not be fed. They need to develop slowly. For this reason you shouldn't apply a nitrogen-rich fertilizer, which would promote immediate leafy growth that would be vulnerable in frosts. Net against birds and caterpillars.

Did you know? The Ancient Greeks and Romans grew and ate cabbages, and the statesman Cato the Elder (234–149 BC) advocated their medicinal use for cleansing and strengthening the body long before their antioxidant properties were discovered by modern scientists.

Organic Tip ✔

Spring cabbage does most of its growing over winter when pests and diseases are not around. The right timing, though, is imperative. The plants need to be large enough to get through winter but small enough not to begin hearting up until spring.

SECRETS OF SUCCESS

- Find a warm, sunny position so that your cabbage plants have the best chance of growing and surviving in hard weather.
- Net to prevent bird and butterfly damage.
- Firm the soil down well after planting and after windy weather. Earthing up the stem helps to prevent wind rock.
- Do not feed your spring cabbage plants.

VARIETIES

'Pixie' AGM
Mid-green, compact cabbage plants producing small, well-hearted heads with few outer leaves.

'Advantage' F1
This variety will provide both spring greens and small- to medium-hearted spring and summer cabbage. Very winter-hardy.

'Duncan' F1 AGM
Dark-green, tight, pointed heads – more of a hearting cabbage.

'Winter Jewel'
Very disease-resistant and slow to mature. Provides spring greens in March but can be eaten like a sweetheart cabbage when more mature.

6 Sow Spinach, Perpetual Spinach and Chard

(late July)

THESE THREE vegetables could provide you with leafy vegetables from late September right through the winter if sown now. Spinach is an annual that is in the ground for 3 months or so. Certain F1 varieties are vigorous enough to germinate quickly and give you perfect spinach throughout autumn; the leaves are lush and never bitter. The best two varieties for sowing now are 'Scenic' and 'Toscane', both of which can be sown right up until late August. Most varieties are only moderately hardy, but 'Triathlon' will often overwinter to provide spring leaves.

Perpetual spinach is a hardy biennial that will stay in the ground for up to 2 years. The leaves are edible and taste like spinach, although the tough mid-ribs and stems should be removed. Swiss chard is closely related, but it has thick, celery-like stems that come in white, yellow, orange or red. The stems have a salty, beetroot flavour and they are cooked separately from the leaves. The crinkled, shiny foliage tastes like spinach.

Swiss chard is extremely hardy and makes an excellent crop for spring and late autumn; you can cut and come again, often when little else is available. The plants need plenty of space and it's usual to leave a 30cm (12in) between each. This crop can look very handsome and colourful. Rhubarb chard has bright-red stems, while Rainbow chard comes in a mixture of colours.

Did you know? Spinach is mentioned in the first known English cookbook, *The Forme of Cury* (1390), where it is referred to as 'spinnedge' and 'spynoches'. In 1533, when Catherine de Medici (1518–89) became Queen of France, she insisted that spinach should be served at every meal. Many French dishes that include spinach are known as 'Florentine', reflecting Catherine's birthplace in Florence.

SECRETS OF SUCCESS

- Sow crops in warm, sunny positions to encourage good germination.
- Sow thinly and then always thin. Leave 7.5cm (3in) gaps between leaf spinach, 22cm (9in) between perpetual spinach, and 30cm (12in) between chard plants.
- All these crops enjoy moisture, so water them well in dry weather.
- Choose F1 spinach varieties with mildew resistance.
- Fertility is also important, so add a general fertilizer when planting.
- Six Swiss chard plants are normally enough. The white-stemmed varieties are hardier than the more colourful ones.

VARIETIES

Spinach 'Toscane' F1 AGM
Rounded, thickly textured, dark-green leaves. This is the most successful spinach for repeat sowings from April until September. Mildew-resistant.

Spinach 'Triathlon' F1 AGM
Large, pointed, yellow-green leaves. A strong variety for cooking.

Spinach 'Scenic' F1 AGM
Bright-green leaves that can be cooked or eaten as baby leaves. Mildew-resistant.

Swiss Chard 'Lucullus' AGM
Light-green, crinkled foliage and ivory-white stems.

Organic Tip ✔

Pick regularly, as this sends a message to your plant to produce more new leaves rather than flowers.

AUGUST

The days are getting shorter and cooler and the evening dews are reviving the garden. This can be misleading and your crops may look fresher than they really are, so if it's dry keep up the watering. Soak rather than sprinkle, because dribbling a hose over the ground is of little use. Water with cans of water and carefully tip a full can on to the ground every other day if you can.

Carry on picking fruit and vegetables and, if you've got too much, give it away to friends or freeze it. If a holiday beckons, ask a neighbour or friend to pick your beans so that they're still productive when you return home. If they're allowed to develop seeds, they'll give up for good.

The weeds will have slowed down, but as soon as there's a note of autumn in the air there will be a rush of weed seedlings popping up so keep up the hoeing and weeding. If it's been a good year you should be in the land of plenty by August and by the end of the month your plot will have taken on a gentle, almost decadent air.

If you've planted an early apple such as 'Worcester Pearmain', you'll be crunching bright-red, strawberry-flavoured apples by now. It's fine to gorge on them because they won't store for long so I never begrudge the starlings, blackbirds, thrushes and other birds a few. You'll also have to watch out for wasps: they switch from an insectivorous diet to a sugar-rich one during this month. It goes without saying that I don't approve of traps, because the wasp is an incredibly good predator for much of the year.

Plums are part of August's bounty and one of the most accommodating fruits for the British climate – although there are good plum years and bad. Some varieties crop reliably, especially 'Victoria', though there are others with a better flavour. If you have room, grow a range of different varieties, because it's possible to pick plums from July until early autumn. If plums fall to the ground, peacock butterflies will dance attendance, but you should remove any mummified fruit stuck to the branches before winter sets in. These can harbour disease.

FRUIT

1 Thin Summer-fruiting Raspberry Canes

(early August)

SUMMER-fruiting raspberries will soon be finishing if they haven't done so already. When they have, it's time to cut down the spent canes at ground level in order to allow the new ones more food and space to develop. Ideally these new canes should be 10cm (4in) apart and any weak canes or very overcrowded ones should be cut out. It's easy to see the difference between the old and new, as this year's will still be light green while last year's now wear a darker patina.

Once your summer-fruiting raspberries have been thinned, feed them with a general-purpose fertilizer and then mulch them with garden compost. You can also use farmyard manure from a reliable source.

Have a good look at the leaves to see whether they are showing signs of chlorosis, or yellowing. If so, the veins will stand out very greenly. It will almost certainly be an iron deficiency and raspberries grown on alkaline soil are especially likely to suffer from this. Apply an iron-rich supplement like Sequestrene 138, using the recommended dose on the bag. If the leaves do not improve, it may be a virus. Unfortunately, raspberries that fall victim to viruses produce stunted canes and little fruit, so prompt action is essential. Dig out affected canes and discard or burn them. Do not put them on the compost heap. There is no other treatment.

If you have room for more canes, growing an autumn-fruiting variety should be an essential because these fruit until late autumn, producing soft fruit when there's little else around.

Did you know? The raspberry is native to almost all parts of Britain except the wet soils of the Fens. It prefers open woodland and scrub, where it often forms extensive thickets. Its seeds are dispersed by birds.

Organic Tip ✔

Summer-fruiting raspberries are more prone to attack by the raspberry beetle than autumn-fruiting varieties. In autumn, dig over your plot to uncover overwintering beetles and larvae so that the birds will find and eat them. If you have chickens, they can help. Mine spend winter in the fruit cage, fertilizing as well as controlling pests.

SECRETS OF SUCCESS

- Once you've cut out the old canes of the summer-fruiting varieties, tie the new ones into the support framework as soon as you can.
- Restrict unwanted canes – they tend to ramble into paths, etc. Chop them out with a spade.

VARIETIES

For varieties of summer-fruiting raspberry, see March, page 86; for varieties of autumn-fruiting raspberry, see February, page 39.

2 Prune Blackcurrants

(early August)

ALL CURRANT bushes make excellent foundation planting in a garden and currants contain a lot of vitamin C. Different types, however, need different treatment. Whereas red- and whitecurrants fruit on older wood and so are pruned as the sap rises (see page 74), black-currants fruit on new wood and are more vigorous growers, so the technique is to remove two or three of the older, darker stems at the base every year. This can be done when picking the fruit, or shortly afterwards, or you could do it in winter; however, it is far easier to separate the new wood from the old now, because the freshly produced wood is a pale brown while the old wood is darker. In winter it can be more of a challenge to spot the age of the branches. I also think pruning now produces a vigorous response and your bush is more likely to produce stronger branches.

A small pruning saw or loppers will do the job. Identify the darkest wood and cut out roughly three branches right at the base. Many gardeners cut off branches laden with fruit because the currants are much easier to pick on a table than by bending down.

You are aiming (with all currant bushes) to create a strong framework of branches that allows the air to circulate. Once the leaves are off it is possible to examine the shape of the bush and then remove any weak shoots, any shoots that are making the middle congested and any branches that are too low down. A short trunk about 20cm (8in) high is ideal.

SECRETS OF SUCCESS

- For advice on growing blackcurrants, see March, page 72.

Did you know? In North America the European blackcurrant is host to a devastating disease of white pines called white pine blister rust. This fungal disease needs both plants to complete its life cycle, so European blackcurrants are banned in states where white pine is economically important. American species of blackcurrant are much less susceptible.

Organic Tip ✓

Blackcurrants are hungry plants and they like a nitrogen-rich feed just as spring breaks – see page 72. They are perfect plants to grow if you keep free-range chickens and they greatly benefit from a winter of chicken manure. Your chickens will have to come out of the fruit cage in spring before the leaves emerge, however.

VARIETIES

For varieties of blackcurrant, see March, page 74.

3 Plant Misted-tip Strawberries

(*mid-August*)

MISTED-TIP strawberry plants are a recent innovation for the home gardener, although they have been used by the trade for many a year. They are raised from unrooted runner tips from the mother plant and then are grown under mist – hence the name. They arrive in bundles with bare roots, so they need to be planted quickly. If you can get them in now you should be able to pick a crop next year. They quickly produce strong, healthy plants and first-year yields are up to 100 per cent better than from other types of strawberry (see page 71). This has to be good news for gardeners.

However, putting plants into the ground in August requires care: young plants need to be kept very well watered throughout this month as the weather can be hot and dry.

Misted-tip plants are available right up until late September, but although this is almost certainly a better time to establish them, it is too late for them to produce a full crop the following summer.

Did you know? The strawberry is neither a true berry nor even a true fruit. The true fruits are the tiny little pips held on the surface of the enlarged, red, succulent receptacle. The receptacle is not part of the flower; it is the tip of the flower stalk on which parts of the flower are held.

Organic Tip ✔

You may see some damage to ripe strawberries caused by a medium-sized black beetle. Do not begrudge the strawberry beetle some fruit – it eats slugs too.

SECRETS OF SUCCESS

- For advice on growing strawberries, see pages 40, 69, 103, 134 and 174.

VARIETIES

For varieties of strawberry, see February, page 42, and March, page 72.

4 Prune and Shape Trained Trees

(*mid-August*)

IF SPACE is in short supply in your garden, restricted trees are the answer. Several shapes are available and they are useful along paths, round edges and close to walls and fences. They mix well with the flower garden and add a formal framework or structure in winter. Although they are unlikely to crop heavily, the fruit that they do produce is often of higher quality than that from full-size trees – and it is easier to pick. Mildew and disease are less of a problem as the air flow is better in a tree that is pruned and trained into a particular shape.

Trained apples and pears are often grafted on less vigorous rootstocks. The strong framework created by training supports short spurs, so all the fruit is close to the framework. Varieties are chosen carefully for their moderate vigour, so the last thing the gardener wants to do is promote a surge of growth. For this reason trained fruit trees are lightly pruned back to their original shape when the tree's sap is beginning to slow down as temperatures cool and day length gets shorter. Make sure you sterilize secateurs between each tree.

Properly trained trees are expensive and, if at all possible, should be personally chosen rather than ordered over the Internet. They are worth the expense because it is a very skilful job to produce a fine fan, or an espalier. Cordons are easier – but still worth buying ready-made. Otherwise you have to spend many years forming the shape.

For the amateur gardener, rubberized Soft Tie wire makes an ideal material for tying in. Simply maintain the tree's original shape. This is made easier by less-vigorous rooting stocks.

Did you know? The word 'espalier' is French and comes from the Italian *spalliera*, meaning something to rest the shoulder against. During the seventeenth century the term referred to the framework supporting the tree, but later it came to be used to describe the tree itself. The practice of growing trees as espaliers became popular in Europe in the Middle Ages, as it allowed fruit to be grown against castle walls and in courtyards.

Organic Tip ✓

Yearly attention is vital with trained trees because an open framework will allow air to circulate. This helps to prevent fungal diseases.

SECRETS OF SUCCESS WITH TRAINED FRUIT

- Buy first-rate trees from a reputable expert fruit nursery and ask them to recommend suitable varieties on the correct rootstocks.
- Use extra-strong galvanized wire and straining bolts for support and for training your tree. Or you could use the self-tensioning Gripple system: this is easier for the amateur to set up.

RESTRICTED TREE SHAPES

Espalier

The most complex way to grow fruit against a wall or on wires. The main trunk has balanced horizontal stems on each side and these are supported on wires. There are usually four well-spaced arms on either side and fruit spurs are encouraged along the length of each.

Cordon

These are usually planted obliquely (i.e. at an angle) because this slows down the sap and makes the tree more fruitful. The fruiting spurs are held along the main stem. Cordons can be planted close together.

Fan

A short trunk supports radiating branches that stretch out to emulate a peacock's tail. Best up against a wall or fence.

Stepover

A low-growing tree for edging paths, supported on a low, strong wire.

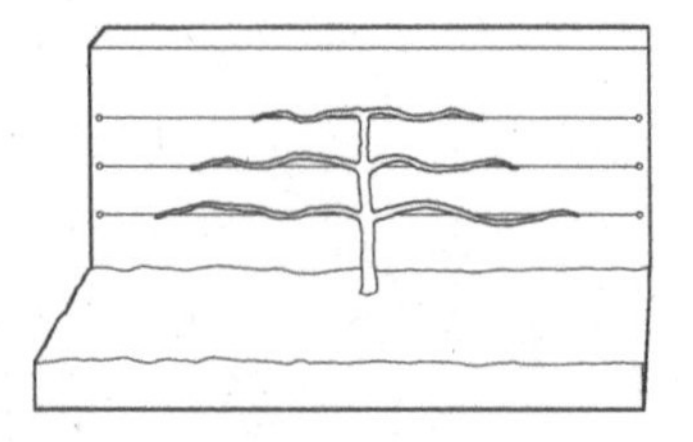

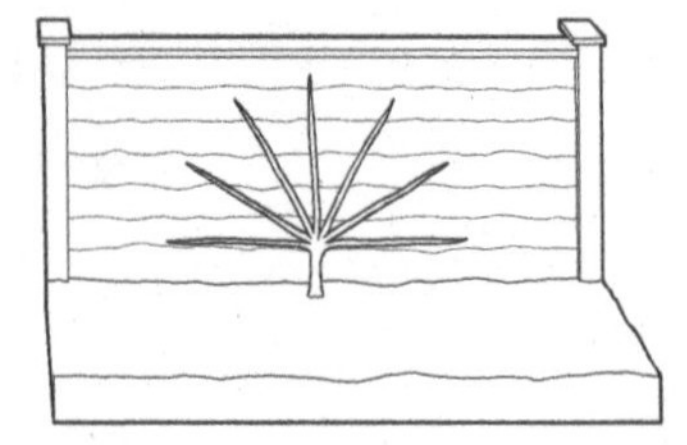

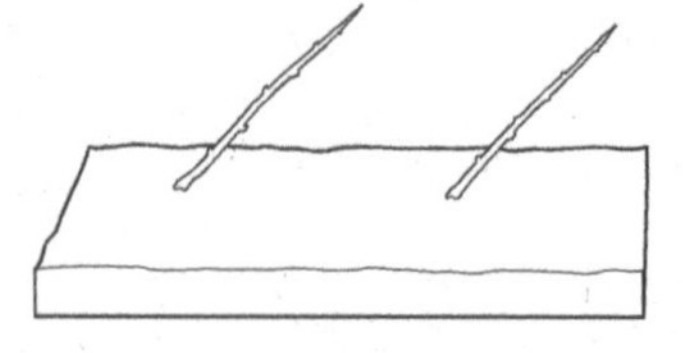

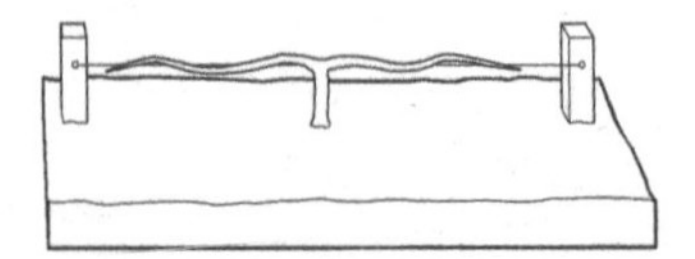

5 Harvest Figs, Cobnuts and Filberts

(late August)

FIGS SHOULD be ripening on the trees now if there has been a good summer. They won't keep, so enjoy them and give any surplus away. Wasps and ants can be a nuisance, so picking in the cool of the early morning, or in the evening, should be less hazardous than during the day. Any figs larger than pea-sized that haven't ripened properly by now should be removed. The pea-sized fruits that remain will develop into your main crop next year, so leave these well alone.

This is also a key time for harvesting hazelnuts – filberts and cobnuts – hopefully before the squirrels descend and strip them from the trees. A good tree can produce 18kg (40lb) of nuts. Try to pick them on a dry day when the husks are just turning yellow. Don't harvest if the husks are still green because the nuts will tend to shrivel in the shell when stored. Lay your harvested nuts out and, once the husks are dry, store them away somewhere cool where rats and squirrels can't get them, such as a garden shed. A hessian or netting sack is ideal. They will keep throughout the winter.

This is a good time to think about ordering nut trees for September or winter planting. They are easy to grow in a variety of soils, except clay. See page 47 for advice on growing hazelnuts.

Eating hazelnuts is very good for you. They provide fibre, protein, vitamin E, minerals and cancer-fighting antioxidants. They also have a high percentage of unsaturated fats – the good fats that help to keep cholesterol levels down.

Did you know? It is said that the evenly spaced hazelnuts in the nuttery encouraged Vita Sackville-West and her husband Harold Nicolson to buy Sissinghurst Castle in Kent. She first came to see it with her son Nigel in April 1930 when looking for an old house where she could make a new garden. Vita fell in love with Sissinghurst and bought it, along with 400 acres of farmland. The garden is now owned by the National Trust.

Organic Tip ✔

Hazel-leaf litter is highly popular with hibernating insects and small mammals because it's light and dry. Leave it in situ *under the trees and it will rot down to produce friable leaf litter, making it a perfect medium for snowdrops and hellebores.*

SECRETS OF SUCCESS

- Cobnuts and filberts will grow in shade, but they fruit better in good light.
- Try to beat the squirrels to your crop: they have a habit of burying nuts in the lawn and this can cause problems as they scrabble to find them.

VARIETIES

For varieties of fig, see October, page 290; for varieties of cobnut and filbert, see February, page 50.

6 Protect Grapes from Wasps

(*late August*)

I COULD never condone wasp traps, because wasps are among the most efficient predators in the garden. Our larger wasps – the yellow-and-black ones – are fantastic predators. Almost nothing preys upon the caterpillars of the Cabbage White butterfly because they are full of noxious chemicals, but wasps do. Caterpillars are attacked and killed, cut up and flown piece by piece back to the hive to provide protein for young wasp grubs. As the summer fades, however, wasps switch to feeding on sugary foods like plums, apples and grapes, so the best way to deter them from eating your fruit is to provide plenty of nectar. They adore red hot pokers (*Kniphofia*). The larger, later-flowering pokers such as 'Prince Igor' produce copious nectar at just the right time. (See also Organic Tip overleaf.)

Grapes are a particular favourite. If wasps find them they will return time and time again. Deter them by tying any fine-mesh muslin, old net curtains or a commercial soft insect mesh around the ripening branches until the fruit is picked.

Did you know? There are 7,500 species of wasp native to Britain and about 6,000 are parasitoid. This word was coined to describe animals that fit somewhere between predators and parasites. Typically, they lay eggs in their host which hatch into larvae, the larvae consume the host from within, killing it, and then they emerge as adults to repeat the cycle. They are tiny – smaller than their hosts, which include many garden pests such as whitefly and aphids. New species of parasitoids are being discovered all the time.

Organic Tip ✔

Some of our British native flowers, such as figwort (Scrophularia nodosa)*, specialize in attracting wasps as pollinators rather than bees. A patch of this in an out-of-the-way place will clear the rest of your garden of wasps.*

VEGETABLE

1 Lift Garlic, Onions and Shallots

(early August)

THE BULBOUS members of the Allium family should all be ready or almost ready to lift. As soon as you see the foliage begin to yellow and flag, it's time to act. If you leave these even 10 days too long they may re-sprout in heavy rain. That makes the bulbs impossible to store, as a green shoot forms in the heart of the bulb.

Shallots should now be fully grown and splaying their bulbs out in a cluster. Onions should also be a good size by now. In sunny summers the ripening process turns the leaves yellow; in wet summers the drying process may need a little encouragement. Lift the bulbs away from the soil with a fork so that the shallow roots break contact with the soil. Leave them in the ground like this for 2 weeks or so and then gather them up. Space them out on a slatted bench or table in full sun. Once the skins are dry and papery, store them in a garden shed.

Garlic should be harvested as soon as the leaves go floppy, which can happen from the end of June until now, depending on when it was planted. The bulbs usually lie underground, so lift them carefully because garlic bruises easily. Shake off the soil and place them somewhere warm and dry before storing.

Organic Tip ✔

The size of your onions is determined by the spaces between them. For a medium-sized onion, space the sets at 10cm (4in). Do not bend the stems over – it encourages neck rot.

Did you know? The Egyptians grew and ate garlic and records from 3200 BC tell us that they regarded it as a sacred plant. Clay models of garlic bulbs were placed in the tomb of Tutankhamun, whether to ward off evil (as it is with vampires) or for its aphrodisiac qualities we shall probably never know. However, aristocratic Greeks and Romans refused to eat garlic, although they fed it to their soldiers and workmen to make them more aggressive.

ONIONS AND GARLICS

Elephant Garlic
This is not true garlic – it is closely related to the leek instead. However, it forms huge bulbs containing four or five large, mild-flavoured cloves which can be roasted. Plant in autumn. Store until Christmas.

Garlic
'Iberian Wight'
Large, flat white garlic with purple stripes from south-west Spain. Plant much deeper than most – to 6cm (2.5in) depth – as this variety pushes upwards. Store until January.

Garlic
'Albigensian Wight'
Large, white, flat-topped garlic from south-west France, eaten by the Cathars in the thirteenth century. Keeps until February – a month longer than other varieties.

Onion
'Sturon' AGM
The best round, golden onion for storage.

For more garlic varieties, see February, page 55.

SECRETS OF SUCCESS

- The drying process is important. In wet summers you may have to dry your bulbs either indoors or in the greenhouse before storing them.
- Golden varieties of onion and shallot store best, and the rounder bulbs are better than the more oval ones.
- Store in the shed in open-weave plastic boxes, or hang strings of onions from a beam in the shed. Ventilation is important.

2 Potato Care

(early August)

AUGUST is the month when potato blight takes hold, encouraged by warm, humid summer weather – technically known as 'Smith periods' (see overleaf). The spores are carried on the air, so the disease sweeps over whole areas very quickly. It constantly mutates and potato varieties that once showed resistance may well succumb in the future. It is a headache for breeders and growers.

The first sign is slightly floppy foliage, followed by yellow marks and brown spotting. As soon as you see it, cut the tops off and bin or destroy them to prevent further spread. Do not add them to the compost heap. Cutting off the foliage quickly will limit the disease and, hopefully, protect the tubers. Infected tubers turn brown and rot. Try not to confuse this with natural yellowing when dying back – look for the brown lesions.

Potatoes are members of the Solanum family and the disease can spread to other members of the family, including outdoor tomatoes and aubergines. For this reason try not to place these two crops close to potatoes.

Wild potato species tend to be blight-resistant, and 'Sarpo' potatoes, bred by the Savari family, are now becoming available. Other varieties show great resistance. Not all get blighted tubers, so hunting out blight-resistant varieties is a good idea. Also, many gardeners who grow only first early and second early varieties avoid the disease altogether because by the time it strikes the potatoes are all out of the ground.

Did you know? A 'Smith period' is a 48-hour period in which the minimum temperature is 10°C (50°F) or more and the relative humidity exceeds 90 per cent for at least 11 hours on each day.

Organic Tip ✔

It is almost impossible to buy an organic potato due to potato blight. On average, a commercial potato crop is sprayed against blight approximately sixteen times, so growing your own is well worth it.

BLIGHT-RESISTANT VARIETIES

'Valor' (maincrop)
Oval, creamy-fleshed potato, for all culinary uses. Very disease-resistant. Also good in dry soil.

'Cara' (maincrop)
The ultimate allotment potato, with very rounded, pink-eyed tubers. A high-yielding, floury potato for every use. The firm tubers store really well, rarely succumbing to blight.

'Lady Balfour' (second early)
Pink-splashed, oval tubers with a firm texture. Heavy-yielding potato for roasting and boiling.

'Sante' (maincrop)
Oval, creamy-fleshed potato for boiling and baking. Resists blight in tubers.

SECRETS OF SUCCESS

- Stick to early or blight-resistant varieties.
- Rotate your crops on a 4-year cycle (see page 315) to lessen the chance of disease.
- Remove all the potatoes from the ground when you harvest and dig up any sleepers that pop up next year – they could be harbouring the disease.
- Don't water the foliage after the middle of July. Water the ground with a leaky hose instead so that the foliage stays dry.
- If you get blight, cut off the tops immediately.

3 Stop Tomatoes

(mid-August)

IN THEIR native land tomatoes are short-lived perennials, but we grow them in the summer months only because they are frost-tender. Consequently, they will carry on flowering, but you'll get much better fruit if you restrict the number of trusses by pinching out the tops or side shoots now.

Six trusses is usually a good number for varieties grown under glass. However, outdoor tomatoes are generally restricted to only three or four trusses. Once 'stopped' the plant puts all its energy into plumping up the fruits and the quality of each tomato improves. You don't want a triffid that flowers into late November: the fruit will be tasteless and your plants will get disease. Go for quality not quantity – a good maxim for all vegetable-growing.

Keep up the fortnightly feeding of liquid potash-rich food (either branded tomato food or free comfrey tea – see page 154) until the end of the month, then begin to restrict the watering slightly. Remove any large leaves that are shielding fruit trusses; that way all the fruit should ripen.

I like all my greenhouse tomatoes to be out of the ground by late October. This gives enough time to enrich the area with compost ready for planting out winter salads in their place. Outdoor tomatoes will also be picked in September and, although they are a gamble, they are still worth growing. The flavour is always more intense. Early-cropping F1 outdoor varieties are more desirable.

Did you know? This red fruit was regarded with great suspicion until the mid-eighteenth century. However, the American president Thomas Jefferson (1743–1826) grew 'tomatoes' in his Monticello garden in 1780 and that is where they first became popular to eat. Heritage American varieties include 'Brandywine' and 'Paragon' – but they tend to be difficult to grow in the cooler parts of Britain.

SECRETS OF SUCCESS

- Get the watering regime right (indoors and out) and try to establish a routine. If tomatoes get too dry and then suffer a deluge, the fruit splits. This attracts disease. Watering in the morning is best.
- Feed religiously with liquid tomato food or comfrey tea (see page 154) once the first fruit appears. Every 2 weeks works well.
- Try to use cans of sun-warmed water in early summer to emulate the steamy riversides of South America where tomatoes grow naturally.
- Put tomatoes outside in June (at the earliest) and give them a warm, sheltered position where afternoon sun strikes. Midday sun can cause stress.
- Select early-cropping varieties for outdoors so that they produce fruit before mid-August when potato blight begins to devastate tomatoes and aubergines.
- Seek out blight-tolerant varieties. Many of these are being introduced from eastern Europe.
- Fleece on cool September nights so that the plants do not get checked by cold weather.

BLIGHT-RESISTANT VARIETIES

'Red Alert'
A bush variety that is early, prolific and tasty – but slugs can be a problem due to its low growth habit.

'Alicante'
This cordon variety normally recommended for growing under glass will also crop outdoors, producing shiny-skinned, medium-sized fruit with a good flavour.

For greenhouse varieties, see May (page 151); for outdoor varieties, see June (page 185).

Organic Tip ✔

Remove the lowest truss or two to keep the fruit well above the ground. This will discourage slugs from eating the ripe fruit.

4 Use Compost to Mulch after Rain

(*mid-August*)

HOPEFULLY your compost heap has been working well all summer long. You should be able to feel the warmth through your fingertips, and the level should certainly have fallen. Try to dig it out now, because the next month or two will provide a glut of green material as crops are harvested and cleared.

Well-rotted compost should look crumbly and dark, and smell sweet. If you have material like this at this time of the year, apply it to the soil surface now, making sure that the ground is damp. It will feed your crops and act as a mulch, keeping moisture in. At this time of year a layer of compost soon disappears into the ground, pulled down by worms. It is important to make sure that the plants you mulch are fully grown, as slugs will home in on any tender babies.

The advantage of emptying your compost bin now is that you create space at a time of year when leafy waste is in plentiful supply and the temperatures are still warm. You can get more compost quickly if you can add materials that contain a good balance of nutrients. Having three compost bins is ideal. One full one can be covered up and left to rot down. One can be ongoing, with material being added every few days. Use the empty bin to turn the full one into, by forking it over into the space. Turning aerates the mixture and brings the well-rotted middle out, allowing the outer edges a better chance to rot down.

Did you know? Composting is an ancient art. The people of Mesopotamia were doing it 1,000 years before Moses was born. There are references to composting in the Talmud, in the Old Testament and in ancient Chinese writings. The Ancient Greeks practised composting, taking straw from animal stalls and burying it in cultivated fields.

Organic Tip ✔

Mulching keeps in the moisture but the ground must be damp. However, as the top layer decomposes it uses up nitrogen from the ground below. So if you are using a material like bark (which rots down very slowly), dust the ground with a nitrogen-rich fertilizer like powdered chicken manure (sold as 6X) before mulching.

SECRETS OF SUCCESS

- Site your compost bin somewhere warm.
- Keep the compost moist and covered – opened-up cardboard boxes are ideal.
- Chop large pieces of green material with a spade or invest in a shredder.
- Feel the heap regularly. The time to turn it is when the heap begins to cool down. Turning is the key to quicker compost.
- Do not add too many grass clippings and always try to rot them partially down first before adding them (see page 183).

- Refrain from adding pernicious weeds such as bindweed and the invasive grass commonly called couch grass or squitch (*Elymus repens*).
- Do not add weeds or flowers that seed heavily (like foxgloves, poppies and aquilegias) to garden compost.
- Aim for a layered sandwich (with airy gaps) and add natural accelerators.
- Do not add cooked food unless you wish to attract rats.
- Dog and cat faeces are too acidic and a health risk.
- Do not add diseased green material to your heap. If it looks blighted or virused, get rid of it. Mildewed material can be added.

GOOD COMPOST BINS

The square wooden bins (1.2 x 1.2 x 1.2 metres/4 x 4 x 4ft) with lift-off fronts are the Rolls Royces of compost bins. Smaller heaps do not warm up effectively.

Plastic bins are easier for people with smaller gardens, where space is at a premium.

Compost tumblers are useful if turning the heap is difficult for you. They can make compost within 6 weeks.

Worm composters can be kept undercover, and the vermi-compost is very rich in nutrients, but the expense of setting one up puts off most large garden owners.

5 Sow Japanese Onions

(late August)

IF YOU want onions by midsummer next year you should sow the seeds of Japanese onions now. These hardy onions (mainly bred in Japan and also known as overwintering onions) will come through bad winters, although they often look very shabby. They need only 12 hours of daylight to trigger bulbing up, so they grow actively from the

March equinox until ready for harvest in June. Conventional onion varieties need 16 hours of daylight before nature tells them to start storing food.

Only keen vegetable gardeners bother to grow Japanese onions because they do not store. They need to be used as you dig them, so one row is enough for most families. At first these varieties were available only in seed form. Seeds are sown *in situ* in shallow drills 23cm (9in) apart and the seedlings thinned out to 10cm (4in) apart in the following spring. You can also sow seeds in modules under glass. Place roughly seven seeds in each and plant out the entire module in the second half of September. However, sets are now available too. These are planted in early October.

Did you know? Thinly sliced onions soaked in a bowl of cold water for 30 minutes taste sweeter and the slices keep for longer.

SECRETS OF SUCCESS

- Timing is everything. Sow Japanese onions in the last third of August. Any earlier and they will bolt in May.
- September is too late to sow: they will not be large enough or well-rooted enough to overwinter.
- Water well after sowing or planting. Water again in March in dry springs as the bulbs need to swell.
- Keep them well weeded.
- The weather will dictate the success of this crop.

Organic Tip ✔

If you suffer from onion diseases like mildew, do not attempt to grow overwintering onions, as you will keep the disease going in your garden throughout the year.

VARIETIES

'Senshyu'
One of the original Japanese overwintering varieties which has proven successful for many. Semi-flat with yellow-brown skin. Matures in June.

'Shakespeare'
A British-bred Japanese/eastern European hybrid producing darker-brown bulbs than most varieties. Harvest from early July. This variety can be stored.

'Radar'
Pale to mid-brown onions from mid-July, although green bulbs can be pulled from the garden from late May. The rounded bulbs have a mild taste and will keep well until autumn.

'Electric'
An overwintering red onion that is becoming popular with vegetable enthusiasts.

6 Sow Green Manure

(*late August*)

GREEN MANURE crops are an effective way of enriching the soil. The most successful are the short-term green manures that germinate quickly and are then dug in over winter. They can be sown now and dug into the ground in 2–3 months in a leafy state. Some long-term varieties (such as alfalfa), which stay in the ground for 2–3 years, have long tap roots that penetrate the ground. Although the deep roots bring nutrients up to the topsoil, these can be extremely hard to get rid of. In any case, most gardeners cannot afford to tie up the ground for this long.

Green manures can be very useful to the gardener because they smother the surface and exclude weeds. On light soil they prevent soil erosion, stop nutrients leaching out of the soil and enrich the soil. If you can dig them in young, they will decompose quickly. However, if they get too tall and the work looks onerous, you can mow the top growth off, allow it to re-sprout a little and then dig it in. This decision will depend on whether you want to use the ground next spring. Bulky material is likely to take a full year to decompose. In August green manures germinate quickly, but as September wears on germination can slow down. If you plant late you risk unwanted seeds popping up in spring. Choosing which green manure to use is vital. Some are more suitable for spring sowing than autumn sowing. Weeds are also a green manure in themselves – but all too often they set seed quickly, so they must be dug in before they flower.

Did you know? Lupins were the original green manure. The Incas used them over 2,000 years ago to protect and enrich the soil and to prevent soil erosion.

Organic Tip ✔

Green manures increase the percentage of organic matter in the soil and improve water retention and soil structure.

SECRETS OF SUCCESS

- Choose your crop carefully. It is the quick crops that are sown now. Long-term green manures tend to be spring-sown.
- Many green manures are nitrogen-fixing legumes. Extra nitrogen can be added by sprinkling blood, fish and bone (see page 317) or by using pelleted chicken manure. Don't discount easier green manures like phacelia. It's the extra organic matter that they add that is most desirable.
- Sow in damp weather for better germination.
- Dig the crop in before it flowers.

GREEN MANURES

Mustard (*Sinapsis alba*)
This is the fastest green manure crop, but it runs to seed quickly. It's frost-prone, so it almost dies off, leaving little work. It is the simplest green manure – although it is a brassica, so don't grow it if you have club root.

Purple Tansy (*Phacelia tanacetifolia*)
This germinates easily in warm soil and produces growth quickly. Once there's a covering, dig it in because in mild winters it will flower and set seed.

Red Clover (*Trifolium incarnatum*)
Sow in autumn for digging in during spring; or sow in spring for digging in during autumn. It sweetens and also helps to lighten heavy soils, and it provides a great deal of organic matter. You need to dig this in thoroughly.

Fenugreek (*Trigonella foenum-graecum*)
This half-hardy annual grows very quickly and produces lots of leaves. When dug in, these enrich the soil.

AUTUMN TASKS

FRUIT

- Cut out the old canes of blackberries and other hybrid berries and tie in new canes.
- Weed round the bases of fruit trees.
- Order new fruit trees, bushes and soft fruit.
- Harvest apples and pears on dry days. Store the fruit in a frost-free place.
- Take cuttings from gooseberries and currants now.
- Apply grease bands to protect against winter moth.
- Check tree ties to make sure they're not chafing the bark.

VEGETABLE

To Do

Explore the depths of the compost heap and empty as much as you can
Cut and ripen squashes and pumpkins
Cut down asparagus as it yellows
Lift all potatoes by the end of September
Weed and clear old crops
Begin winter digging
Collect leaves
Check nets on brassicas – hungry pigeons are about

....

Sow under Glass

Oriental salad leaves, radicchio and endive for winter use
Spinach

....

Sow Outside

'Aquadulce Claudia' broad beans (early November)
'Feltham First' peas (early November)

....

Plant

Elephant garlic and autumn-sown varieties of garlic
Transplant spring cabbages

SEPTEMBER

September is probably the most mellow month of all, and when the apples and pears begin to colour up and ripen, their sweet aroma wafts through the garden on warm afternoons. One of the great delights of gardening is to gently twist a sun-warmed, perfectly ripe apple and then sink your teeth into it. It satisfies a primeval instinct.

Gardeners do not have to trouble themselves about whether the fruit is firm enough to travel hundreds of miles in a lorry, or whether it's perfectly shaped, or large enough to tempt a buyer. The odd blemish does not matter either, and the fact that my apples are not chemically tainted is highly important to me. The only consideration is flavour, and my 'Pitmaston Pineapple' produces small, yellowish apples with an aromatic pineapple flavour. 'D'Arcy Spice' is even better, with its hints of Christmas spice and cinnamon.

There should be a glut in the vegetable garden this month and you should be harvesting beans galore, picking courgettes on an almost daily basis and harvesting peppers, tomatoes and cucumbers at least twice a week. The whole garden wears a contented air, and crystal-clear September light, produced by evenly balanced days, flatters everything.

On a cautious note, September frequently brings the first frost and there may be a really long gap before you get another. Be prepared to fleece your courgettes, winter squashes and tender beans if a cold night is forecast; this will keep them going for several more weeks.

FRUIT

1 Harvest and Store Apples

(early September)

GROWING apples should provide a crop that can be stored and, if you intend to do so, they will have to be picked from the tree before they crash to the ground and bruise. They must be full-sized and ripe, and the time-honoured way to tell this is to cup them and give them a twist. Most ripe apples will happily detach themselves from the stem. However, later varieties often need encouragement in the form of a sharp tug – like intransigent children. They are descended from wild genetic material that clings on to its fruit until the following spring, so these varieties rarely detach easily.

Different apples store for different lengths of time, so mark the variety (if you know it) and then add the words 'eat by' to each basketful. For this reason varieties should not be mixed up: if they are, some will rot and spoil the other, longer-lasting types. Apples give off ethylene, a gas, when ripe and this causes fruit nearby to ripen prematurely. Green bananas or an under-ripe avocado placed next to your apples will ripen quickly, so for this reason bananas are normally kept away from apples to prevent premature ripening.

Generally, early-ripening apples keep for 3 weeks at most and are probably best left on a wooden tray somewhere cool. If it's a bumper crop, give some away. Mid-season-ripening fruit keeps for 1–2 months. Late-season fruit (picked in October) will store for the longest of all – between 3 and 8 months. It obviously makes sense to grow at least two late-maturing varieties that can be eaten after

Christmas. Don't eat these later apples straight from the tree: they have to be stored to develop flavour and sweetness.

Did you know? Fruit stores traditionally had a soil floor which added a touch of humidity that kept the fruit from shrivelling. A layer of strong mesh wire was usually laid underneath the floor to prevent mice and rats from getting at the fruit. If your store is dry, you may have to dampen it a little by putting out a tray of water.

SECRETS OF SUCCESS

- Apples must be perfectly dry before storage, so pick in the middle of a fine day.
- Look carefully for any breaks in the skin or signs of insect attack. If you're sure the apples are completely sound, carefully place them in a basket (handling them just as you would eggs).
- If you have room, leave them in a ventilated place for 10 days before wrapping, in case they sweat.
- Wrap them in paper or put them in cardboard apple trays, which you can usually acquire from a supermarket or fruit stall.
- Wooden orchard trays stack on top of each other and allow the air to circulate, which prevents mould, etc. But an old chest of drawers (with each drawer left slightly open) works too.
- Apples are always stored on their own, well away from pears and other fruit. They pick up taints easily, so you must move strong-smelling tins containing creosote, etc., well away from them.
- A cool, rodent-free garden shed is ideal. Keep the door shut, otherwise birds will soon peck your fruit.
- Check your fruit every month and remove any that show signs of rotting.

Organic Tip ✔

Wrapping stored apples, to ensure the apples never touch each other, prevents a rotten one from infecting the whole batch. The paper barrier keeps the mould spores from escaping. Waxed paper is the ideal, although newspaper has been used for generations. Leave a gap between each wrapped apple. Alternatively use see-through polythene bags so that it is easy to check the apples regularly, removing any that show signs of rot; each bag should store up to 1.8kg (4lb). Make holes in the bags to allow for ventilation. A garage, cellar or outhouse is the perfect place to keep them.

VARIETIES WORTH STORING

'Ashmead's Kernel'
A classic, russet-coloured old English variety. Excellent aromatic flavour, good keeping quality and very attractive blossom. Eat between December and February. Raised in Gloucestershire, *c.*1720. Pollination Group D.

'Bramley Seedling'
Waxy-skinned, large, green cooking apple that cooks to a fluffy purée. Stores really well until March. From Nottinghamshire, 1809 (see page 20). Pollination Group C. A triploid tree with sterile pollen, so grow two more Group C trees to go with it.

'Kidd's Orange Red'
Similar to 'Cox', but redder in colour, sweeter in taste and easier to grow. Eaten between November and January. From New Zealand, 1924. Pollination Group C.

'Blenheim Orange'
A nutty-flavoured old English variety producing large orange fruit streaked in red that can be eaten or cooked. Heavy-yielding with a biennial tendency that is mostly overcome if grown on dwarfing rootstocks. A big triploid tree, so you will need to plant it with two other Group C trees. From Oxfordshire, 1740. Pollination Group C.

For further varieties of apple, see January, page 14, and September, page 263.

2 Harvest Pears

(early September)

PEARS RIPEN rapidly and are entirely different beasts from apples. They don't crop as heavily and they don't store nearly as well. Picked ripe, many develop a 'sleepy' middle, with the centre quickly turning to brown mush. Pears, therefore, are best picked from the tree when ripe: then they will keep in the bottom of a refrigerator (which should stay just above freezing) until November. The one advantage that pears do have over apples is that they can be stored in much cooler temperatures.

When ripe, pears change to a lighter shade of green – that is the time to pick them. They will come away from the stalk and the fruit will have a lovely aroma.

Some heritage varieties can be stored until January, but you need to pick these before they are fully ripe, so check your varieties before planting if you want to store.

VARIETIES FOR STORING

'Passe Crassanne'
A heritage French dessert variety from 1845 with a buttery texture, requiring a warm location – often better against a wall. Stores until February or March. Pollination Group B.

'Packham's Triumph'
An Australian variety from 1896 – effectively a late 'Williams' with yellow, musky flesh. Stores until November. Pollination Group B.

'Winter Nélis'
Not a strong grower, so needs grafting on to Quince A and the small, russet-skinned fruits do require thinning. However, stores until January. From Belgium, 1818. Pollination Group D.

'Glou Morceau'
This buttery 1759 variety from Belgium is the longest-storing pear. Consume between December and January. Crops well, but enjoys sun and warmth. Pollination Group D.

For further varieties of pear, see January, page 18.

Did you know? The most famous British pear variety is the 'Williams', but it wasn't deliberately bred. It arose as a seedling in the garden of a schoolmaster, one Mr Wheeler of Aldermaston in Berkshire, sometime before 1770. It was originally grafted by Richard Williams, a nurseryman from Turnham Green near London, and it took his name. In fact, its correct name is 'Williams Bon Chrétien', meaning 'Williams good Christian'. It is the most widely grown pear in California (where it is known as 'Bartlett's William') and it is the one most often canned.

Organic Tip ✔

Heritage varieties of pear are especially fond of warmth and they should be given the warmest spot possible. Their blossom is frost-hardy, but they need the extra warmth for the pollen tube to grow.

SECRETS OF SUCCESS

- Be realistic! Pears do not store nearly as well as apples.
- Fully ripe pears can be kept in the bottom of the fridge in polythene bags for up to 6 weeks.
- If storing for longer, pick your fruit before it is fully ripe and keep it as cold as possible.
- Lay pears out on wooden or cardboard trays, variety by variety, in an airy place.

3 Grow Later Varieties of Apple

(mid-September)

IF YOU are serious about storing fruit, it's the later varieties of apple that store best. These also tend to flower later and many are good varieties for colder gardens, as the blossom generally misses the frost.

You will still need to avoid frost pockets and plant them in good light so that nectar flow attracts the bees. Plant bare-root trees between November and March; pot-grown trees can be planted throughout the year, but avoid planting in extreme cold or extreme heat. To fruit well apples, pears and plums need cold winters so that they have a full period of dormancy. Given this, they form big, bold, efficient fruit buds. However, the number of chilling hours needed varies according to variety. Sunnier summers also promote bigger fruit buds. The ideal would be a hard winter followed by a warm spring and a sunny summer.

Organic Tip ✔

Ornamental crab apples make excellent pollinators for apples. Possible varieties include 'Golden Hornet', 'Red Sentinel', 'Winter Hornet' or 'Profusion'.

SECRETS OF SUCCESS

- For advice on harvesting and storing apples, see page 257.

Did you know? Summers are getting hotter in England, particularly in the south-east. Because English apples tend to be thin-skinned, with low wax levels on the skin, the fruit can suffer sunburn and can almost cook on the tree. If hot summers persist, varieties with thicker, waxier skins will have to be grown instead. 'Braeburn', bred in New Zealand in 1952, is now widely grown in England for this reason.

RECENT VARIETIES THAT STORE WELL

'Jonagold'
Heavy-yielding apple from the USA, 1943. Fruit is greenish-yellow, streaked in red. Crisp and sweet, with a rich, honey flavour. Stores November–January. Pollination Group D.

'Red Falstaff'
Late, red dessert apple from Kent, 1986. Stores October–December. Pollination Group C.

'Rajka'
Disease-resistant, mid–late red dessert apple from the Czech Republic, 1990s. Stores October–December. Self-sterile. Pollination Group D.

'Pinova'
Late, fruity dessert apple with good disease-resistance from Germany, 1986. Stores November–January. Pollination Group C.

4 Harvest Quinces and Medlars

(mid-September)

QUINCES and medlars, both popular fruits for centuries but grown less often today, can also be harvested now. Medlars (*Mespilus germanica*) are not universally popular because the fruit has to 'blet', which means that you eat it in a brown, almost rotten state. It is an acquired taste and the fruits are small. They do, however, make excellent jelly.

Medlars also make good specimen trees for the ornamental garden because their white single flowers on the tips of the branches have a pure charm and they come after the apple blossom, thus extending the blossom season. Bright-green leaves follow and then the fruit forms. Pick them now and store them for a few weeks. They are inedible when fresh, so lay them out flat and leave them in a warm place for 4 weeks or so. They become soft and brown and the flesh can be spooned out. Rely on your tastebuds and nose if you leave them longer.

The quince (*Cydonia oblonga*) is really in the same category and mainly used for making jelly. Quinces hang like huge, ugly pears and they are also picked now and left in a bowl to soften a little. They take ages to cook. At blossom time they are extremely ornamental, with huge, apple-blossom-pink flowers that open from scrolled buds. It is my favourite blossom of all. Sometimes the flowers are veined in darker shocking-pink. They love to grow in gardens close to rivers or in the rain-sodden counties of western England.

Both quinces and medlars come from Asia and so will do best in warmer gardens. Both are self-fertile, so you need only one of each. Different rootstocks are available for both. Trees grafted on to Quince C can be kept small, with blossom appearing at eye level.

Organic Tip ✓

All fruit trees are important sources of pollen and nectar for our endangered bees. A quince, or a medlar, extends the flowering season and the bees flock to them.

Did you know? The medlar is commonly known as 'dog's bottom' or 'monkey's bottom' and you can see why when you look. Other descriptions are even ruder. The tree originated in Persia, becoming popular in Britain in the Middle Ages. The 'bletted' fruit was traditionally eaten at Christmas.

SECRETS OF SUCCESS WITH QUINCES AND MEDLARS

- A warm, sunny position is vital for medlars and quinces.
- Quinces need moisture to produce those huge fruits.
- Opinion varies here, but I prefer to harvest medlars and quinces before the first frosts and store them in a warm place.

VARIETIES

Medlar
'Nottingham'
A very ornamental variety with a good architectural shape that sends branches in odd directions. The foliage colours well in autumn.

Quince
'Vranja'
Grown for its exceptional flavour and perfume. Produces large, golden-yellow fruit. A larger tree than the medlar 'Nottingham'.

5 Pick and Plant Walnuts

(late September)

WALNUTS are so nutritious and high in vitamin E and the 'good' fats that lessen cholesterol that every gardener with room and a suitable site should aspire to grow them. With luck some nuts will survive the annual invasion of agile squirrels swinging Tarzan-like through the branches. Walnuts are not normally picked, but gathered from the ground on a daily basis. The husks, which usually split at this stage, need to be removed because they blacken and go mouldy if left. They can also taint the nut inside. Stored nuts are then kept and dried out so that they lose their wet texture and develop their oily flavour. They are ready to eat when dry to the touch. This usually takes a few weeks.

Some people like to pickle walnuts, but for that they must be picked from the tree while very young, before the shells have begun to develop – usually in July or August.

It is important when starting out to buy a young grafted tree. This will fruit after 4 years, whereas an ungrafted tree could take 20. Plant it in the autumn or during the winter rather than in the spring. British springs and summers can be dry and the walnut is used to an Asian rainy season that produces a heavy summer deluge in the growing period. Spring-planted walnuts seem to get a growth check unless there's a wet summer.

If you have room plant at least two, because although some varieties are self-fertile, the yield is greater if cross-pollination occurs. Walnuts are wind-pollinated, but even compact forms need to be up to 10m (30ft) apart.

Did you know? Walnut husks stain the skin and clothing, so wear old clothes and gloves when handling them. Hard pruning is not tolerated and regular pruning is not necessary. However, walnut trees bleed freely, so if you do need to do it, tackle the job between midsummer and early autumn when the sap is running slowly.

Organic Tip ✔

Walnuts are warm-position trees from Asia and they do not crop well in more northerly latitudes of Britain. Before you plant one, do some local research.

SECRETS OF SUCCESS WITH WALNUTS

- Find a warm, sunny position away from frost. Both flowers and the growth tips can be spoiled by frost.
- Provide deep, fertile soil – heavy loam over limestone is ideal. Avoid water-logged positions.
- Handle the roots carefully when planting and never plant a pot-bound tree.
- Stake your tree when you plant it.
- Keep your tree well watered in dry springs and summers until it is established.
- Don't prune your tree.
- Apply a balanced fertilizer in late February and again in late March. Then add a mulch of well-rotted compost around the base of the tree to retain moisture.

VARIETIES

'Broadview'
This Canadian variety has become the best all-round walnut for UK conditions. It is suitable for smaller gardens and forms a compact, spreading tree. It produces nuts after 3–4 years. The flowers are also quite frost-hardy and it flowers later than most.

'Franquette'
Raised in the heart of Europe's finest nut-growing region around Grenoble in France, this grafted tree produces sweet, moist nuts with a high oil content. Cropping should begin within as few as 3–4 years.

VEGETABLE

1 Sow Winter Salads

(early September)

YEARS AGO salads were purely summer affairs, but in recent years cold-tolerant leaves have become a garden staple. Plant them in an unheated greenhouse or frame and they will produce a crop from October right through to next April, and they are very happy to follow on after tomatoes. Oriental vegetables (like mizuna and pak choi) thrive in cool temperatures and you can plant them alongside rocket, endive, chicory and winter lettuces. Seeds are sown now and young plants should go out in the second half of September or in October (after lifting tomatoes). By then the temperatures are much lower, so there are few slugs and other pests about. The low temperatures also ensure slow growth. Crops that are prone to bolt (run to seed) in summer, such as rocket, will grow on for months, producing leaf after leaf without flowering. And there's no flea beetle either!

Seed companies sell mixtures that vary in flavour from hot and spicy to mild, but you can also create your own mixtures. The easiest technique is to use modules for each variety and place two or three seeds in each. This will save pricking out. Bed out the plant clusters in rows, leaving 10cm (4in) between them. This creates a carpet of leaves, and the textures and colours will vary pleasingly from rounded, deep-green baby spinach to ferny red mizuna and peppery rocket.

Organic Tip ✔

Place upturned plant saucers among your leaves. Slugs will hide under them, making it easy for you to find them. They are especially partial to chicory leaves.

Did you know? Bagged supermarket salads have been found to contain potentially harmful bacteria like *E. coli*.

SECRETS OF SUCCESS

- Sprinkle on blood, fish and bone (see page 317) once the plot is clear. This will boost the soil's nitrogen content.
- Once the plants are in the ground, water them thoroughly every other morning so that they race away.
- Once the plants reach a height of 10cm (4in), pick them regularly to encourage more leaves.
- If you spot any plants running to seed, pick out the tops.

VARIETIES

'Niche Oriental Mixed'
Golden and red mustards, mizuna, komatsuna and rocket. Mizuna is almost fern-like in leaf and komatsuna resembles small-leaved spinach. Both add great flavour and texture as well as looking pretty on the plate.

Rocket
'Apollo'
Peppery tender leaves of just the right size.

Chicory
'Radicchio di Treviso'
Upright dark-red leaves with white mid-ribs – a bitter flavour.

Chicory
'Grumolo Verde'
Rosette-forming, green-leaved chicory that tends to come into its own as the days lengthen.

2 Repair Grass Paths, etc.

(early September)

GRASS PATHS are more environmentally friendly and kinder to ground-hugging friendly predators such as ground and rove beetles. This is the best time to tackle grass repairs – whether it's seeding the gaps in paths or dealing with the weeds. The temperatures are ideal now – not too hot and not too cold. Night-time dew encourages germination and autumn rains boost growth because the soil is still warm. Spring, the other opportunity to sow grass seed, is chancier as the night-time temperatures can be low and spring droughts often occur.

The first thing to do is to remove all perennial weeds, especially dandelion roots; there are specialist tools for this job called daisy grubbers. Distress the bare patches with a short-tined rake or fork, treading it down very lightly before scattering the seeds. Water well, cover with wire netting and keep off the area for 6 weeks. The ground should be kept watered throughout this time, but shouldn't get waterlogged.

Larger areas will need better preparation. Rotavate the area if possible, or dig it through by hand, removing any stones and all weeds. Either add organic material to poor soil, or add grit or sand to heavy clay. Leave the ground to settle for 10 days, then hoe off any emerging weeds. Sow the seeds as instructed on the packet.

Lawn seed mostly contains two main types of grass. Rye grass is tough and fast growing, and therefore quick to establish. Fescues are finer and slower growing, and best for ornamental 'bowling green' lawns. The best option for paths is a mixture containing both.

Organic Tip ✔

Don't use herbicides near vegetables. Spend a few minutes now and again weeding. Dandelions are the most invasive and there are two flushes – May and September. They can be dug out easily now.

VARIETIES IN LAWN MIXTURES

Perennial Rye Grass (*Lolium perrene*)
Rye grass is tough and durable and is found in many mixtures. It germinates within 10 days and forms a dense lawn. It dislikes shade and isn't drought-tolerant, but quickly bounces back after rain.

Fescue (*Festuca species*)
Several different fescues can be included in mixtures. These fine-leaved grasses thrive in well-drained soil in the wild. They form the basis of fine lawn mixtures, but they are also contained in family lawn mixtures to a lesser degree. They are slower growing and tend to stay green in dry weather.

Browntop Bent (*Agrostis species*)
This fine green grass is added to mixtures designed to thrive in shady areas.

Smooth-Stalked Meadow Grass (*Poa pratensis*)
Another grass for shade, with drought tolerance. This grey-green grass may take 21 days to come up.

Did you know? The lawnmower was invented in 1830 by Edwin Beard Budding, an engineer from Stroud in Gloucestershire. He saw a machine trimming cloth in a local weaving mill and realized that it could cut grass in the same way.

SECRETS OF SUCCESS

- Choose the right mixture. One recommended for everyday family use makes an ideal path between vegetable plots. Make sure it's fresh.
- Catch the weather and sow after rain if possible, otherwise water first.
- Don't be mean with the seeds, and rake through after sowing so that the seeds are in close contact with the soil.
- Cover with wire and then fleece to keep the warmth in. Leave the fleece down for a week.
- If germination fails after 20 days, sow again quickly.

3 Harvest Sweetcorn

(mid-September)

ONE CROP that breeders have revolutionized in recent years is sweetcorn. Modern F1 varieties now crop reliably in Britain even in colder areas if planted out in June and watered well. Many of the best British-bred varieties are named after birds. They include 'Lark', a mid-season variety; 'Kite', a later variety; and 'Lapwing', which comes between the two. These extra-tender varieties are almost fibreless: they will not fill your teeth with annoying bits and the bright-yellow cobs cook quickly. Of the three, 'Lark' is particularly recommended for colder gardens and every cob seems to be well filled.

Cobs of corn are very sweet. However, after picking the sugars turn to starch quickly, so always eat your cobs straight away. Supersweet varieties are not necessarily sweeter, but they take longer to lose their sugars. Consequently they are favoured by some commercial growers because the cobs can be stored for longer. Home gardeners should stick to the extra-tender varieties and 'Lark' F1 is the best at the moment. Once the beards are brown the cobs are ripe, but if in doubt pinch a kernel and it will yield a milky fluid if ripe.

Most varieties yield one or two cobs per plant and these should be harvested when ripe. Lift sweetcorn plants and compost as soon as possible.

Did you know? In Latin America sweetcorn is traditionally eaten with beans. Each plant is deficient in an essential amino acid, but it is abundant in the other. Together they form a balanced diet.

Organic Tip ✔

Sweetcorn foliage and stems are cellulose-rich and take a long time to rot down. Snip them into pieces as you add them to the compost heap – this helps them to decompose more quickly.

SECRETS OF SUCCESS

- Sweetcorn is wind-pollinated, so grow only one variety to prevent cross-fertilization.
- Plant in a block (not a line) to aid pollination.
- The secret of a good crop is watering in June and July to create lots of foliage and height.
- If you have a glut, don't store cobs. Cook and freeze the surplus instead.

VARIETIES

For varieties, see April, page 123.

4 The Autumn Tidy

(mid-September)

COOLER nights followed by warm days are a recipe for dewy mornings and evenings. These conditions are perfect for the re-emergence of the slug and snail. These are not all bad. The larger slugs do eat debris, but it makes sense to begin to tidy up the vegetable garden so that there are no hiding places. By mid-September the garden has begun to move towards decay and decline, and many crops are coming to an end.

Frosts are not far away and the slightest one will reduce cucurbits to mush, so if you have any courgettes that look well past their best, or any cucumbers at the end of their productive life, it's better to compost them now before they become slimy to handle. Nasturtiums and rhubarb leaves also go to mush, so tidy the crowns of rhubarb and bin any nasturtiums. Tidy up the brassica bed and remove any fallen yellowing leaves.

Weeds will also be germinating apace and hoeing this month will disrupt them, as well as disturbing any slug and snail eggs in your soil. Do a thorough job now and it will save you hours next spring because weeds develop seeds within a short space of time.

SECRETS OF SUCCESS

- The gullies around vegetable beds (where the lawn or path meets the soil) are seed repositories for weeds and shelter belts for slugs. Clean them out now. Trim the grass and re-cut the edges. It will prevent a lot of problems.
- Tidy rhubarb crowns and compost cucurbits before frost makes them flop and flag.
- Hoe your vegetable beds to keep down the weeds. You will also disturb slug eggs – the birds love them.
- Collect leaves and put them into a wire frame or punctured black plastic sacks. They will slowly form leaf compost, which makes a good mulch.

USEFUL TOOLS

Rubber Rake
Much better and friendlier than a noisy leaf vacuum, this wide rake has rubber tines that can be dragged through plants without damaging them. It's a wizard at collecting leaves from the soil.

Small Onion Hoe
A vital tool for getting between onions or brassicas. The one-sided blade is less dangerous than the swoe.

Swoe
A stainless-steel blade with a double edge that cuts as you go backwards and forwards. This is an excellent tool for weeding bare plots, where you can push it up and down without fear of decapitating plants. Leave the weeds on the soil surface if they aren't flowering and then dig them in as a green manure.

Wheelbarrow
Choose something light, durable and not too garish, with a strong pneumatic tyre and good handles. As with gloves and hats, you have to try them out.

Organic Tip ✔

Don't use slug bait. The big black garden slug (Arion ater) *eats rotting debris, so it helps clean the garden up.*

Did you know? Snails hibernate in sealed shells and cluster together in sheltered places from October to March. However, slugs are active whenever the temperature is above 5°C (41°F).

5 To Dig or Not to Dig?

(late September)

THERE ARE gardeners who dig and gardeners who don't, and there are good reasons for both. No-dig gardeners believe that nutrients should be added to the surface (in the form of a top-dressing) and pulled down by worms. This definitely leads to warmer soil early in the year, because you are not exposing it to the cold, and fewer weed seeds germinate in spring. No-dig gardeners feel their approach keeps nutrients in the ground. The no-dig method is not labour-free, however, and you have to be prepared to barrow in organic matter to cover the plots with a 5cm (2in) layer once or twice a year.

Most gardeners enjoy digging and believe it aerates the soil and improves soil structure. If you are a digger by nature, this is the best time to start – just as the first frosts arrive. The technique is to get a good fork and turn the soil over once. The clods of soil should not be broken down into a fine tilth – they should be left on the surface in large lumps. This increases the surface area of the soil exposed to the weather. The action of freezing and thawing (brought about by frosts) breaks the clumps down for you over winter. All that the gardener needs to do is rake the soil into a tilth in spring.

Did you know? Ruth Stout (1884–1980), who gardened in Kansas, was a famous no-dig pioneer. She advocated using a thick, 20cm (8in) mulch of hay to suppress weeds and keep the soil moist. Stout used cheap 'spoiled' hay that wasn't suitable for animal use. When she planted potatoes she chitted them and threw them on to the surface, and she planted seeds in the same easy way. She also became famous for not watering her garden for 35 years – and Kansas is a dry place in summer.

SECRETS OF SUCCESS FOR A NO-DIG APPROACH

- Top dress with well-made garden compost or well-rotted manure.
- Apply it thinly, aiming to cover to a depth of 5cm (2in).
- Remove perennial weeds (like docks) with a trowel.
- Kill couch grass, dandelions and buttercups by covering them with cardboard.
- Hoe annual weeds in spring when they germinate.
- Root crops like potatoes and parsnips are dug out of the ground, so there is always some soil disturbance even in a no-dig system.
- Raise your vegetable plants in modules in the greenhouse, then plant them out into the garden at the optimum moment so that they get the best start.

Organic Tip ✔

Whenever you use organic matter, whether it's garden compost or manure, it must be well rotted, otherwise it will rob your soil of nitrogen as it decomposes and will actually produce a poorer crop initially. The captured nitrogen is returned to the soil a season or two later.

6 Lift All Potatoes
(late September)

ALL YOUR potatoes should be lifted before the end of this month so that they don't attract slugs, which will pepper any tubers left in the ground with unattractive holes. The potato plant, sensing the underground attack, fights back by making the potatoes less palatable. Affected tubers develop an unpleasant, earthy aroma which you can taste and smell. This can make them almost inedible.

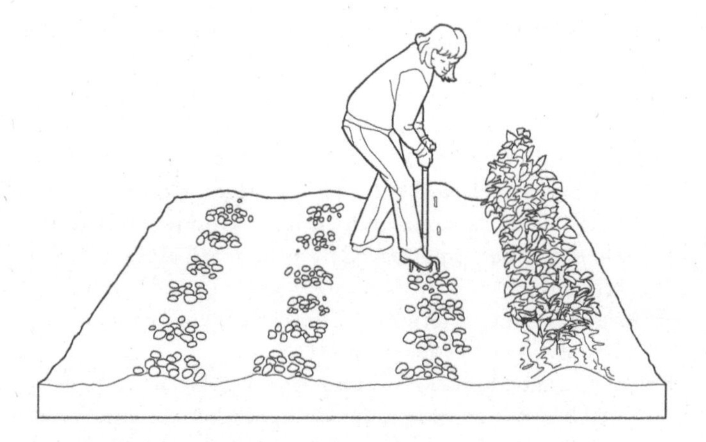

Cut the potato foliage off and allow the ground to dry. Try to find a dry morning to lift the tubers and leave them for a few hours to dry off in the air. Store them in thick paper sacks to keep out the light. If sunlight gets to the tubers, they produce chlorophyll and turn green. As they do so, they produce a bitter neurotoxin called solanine as another defence mechanism. Green potatoes look unpalatable and they can upset the stomach. Some potatoes also produce tomato-like fruits and these are toxic, so be sure to dispose of any lying on the ground.

Store only firm, healthy-looking tubers. If any have marks or dark patches on their skin they may be suffering from potato blight. Discard them. If you've grown different varieties, separate them out into different labelled bags because this makes cooking easier. If you have several varieties they will cook at different rates, whether boiled or baked.

Storage times vary according to type. First early and second early varieties generally do not store well, so use these as quickly as you can. Maincrops should store until after Christmas and possibly into March. However, the tubers are frost-prone, so you will need to find a cold but frost-free place to store them, ideally with temperatures around 5°C (41°F).

SECRETS OF SUCCESS

- Thick paper sacks are best. They should be only half-filled. This makes it easier to tip them out and check for bad potatoes. Often the smell alone will tell you.
- Store different varieties in different bags for ease of use.
- Check all stored crops regularly.
- Aim to use up home-grown potatoes by March, as they will begin sprouting as the days lengthen. Commercial crops are sprayed with an anti-sprouting chemical.

Did you know? Each cubic metre of soil contains on average up to 200 slugs, but the amount of rainfall and your type of soil affects how many you may have. They breed all year round, but egg-laying peaks in March–April and September–October.

Organic Tip ✔

Choose your varieties wisely: some are much less attractive to slugs and many of these types are also resistant to eel-worm. If you find slugs have been wreaking havoc with your crop, treat the ground with slug nematodes quickly whilst the soil is still warm enough *(see page 127)**. This will prevent the underground army moving on to other crops.*

SLUG-RESISTANT VARIETIES

'Anya'
A knobbly salad potato that can be treated as a second early or left in the ground.

'Cara'
This allotment maincrop toughie seems to hold little attraction for the slug.

'King Edward'
An early maincrop with high slug resistance.

'Sante'
An early maincrop widely grown organically in Europe due to its slug and disease resistance.

'Pentland Dell'
Slug-resistant heavy cropper producing white potatoes.

'Kestrel'
A handsome blue-eyed potato often seen on the show bench. Also slug resistant.

OCTOBER

October days are often misty and crisp and there's still fruit to be picked on sunny days, particularly apples and pears. The acid test is to carefully cup the fruit with your hand and gently twist it a quarter turn. If the apple or pear comes away from the tree cleanly, the fruit is ripe enough to pick. Some varieties need storing so that they sweeten up before you eat them. You may well be picking huge autumn-fruiting raspberries, too, and these are the easiest crop of all to grow.

In the vegetable garden, crops often continue to grow until November, so it's worth fleecing slightly tender vegetables if a frost is forecast, using horticultural fleece or double sheets of newspaper. Remove any covering the following morning so that any cold air can escape. If the weather stays mild, later sowings of beans, courgettes, carrots and peas will continue to crop. Winter brassicas, such as Tuscan kale, can also be picked now.

The year is winding down and the leaves are turning warm shades of yellow and orange, before falling. On the work front there is little to be done this month, although October afternoons can still be generously warm. Enjoy it while you can, for we are on the cusp of great changes. In a few weeks' time winter will be upon us, and we'll either be leaf gathering or sitting indoors.

As crops finish, tidy up and remove them and place the debris on the compost heap. Put garden paraphernalia away too. The wigwams of bean canes, for instance, or spent containers of ornamental plants. This will make the garden look better and save you time later. A tidy plot will lessen the chance of fungal diseases, and the gardeners' enemy, the slug, will have fewer places to hide.

FRUIT

1 Prepare Ground for Bare-root Fruit

(early October)

THERE ARE several advantages to buying a field-grown bare-root tree or bush rather than a container-grown one, including the fact that a bare-root specimen is often cheaper. Also, the range of varieties is greater and some rarer types can be bought only in this way. If you've ordered bare-root fruit trees or bushes, use one of these mellow October days to prepare and enrich the ground so that when they do arrive the chore of planting them on a raw day is all the quicker.

Dig over the ground and weed it well, then add organic matter. This could be garden compost or well-rotted manure from a reliable source. However, all manure must smell sweet (not of ammonia) and be well rotted, otherwise it will scorch the roots. There have been widespread problems with contamination of manure caused by a herbicide called aminopyralid, which has been used on grass fields that have then been cut for silage. When this is fed to cows, the manure coming out of the other end is effectively a herbicide too. So if you do use manure, check to make sure that your farmer hasn't used either the herbicide itself or bought-in silage. If he has, avoid the manure.

Once the ground is fully prepared, cover it up with a double layer of horticultural fleece or old carpet; this will keep the cold and frost out of the soil. When your plants do arrive you will be able to drop them into the ground easily.

Did you know? Bare-root planting is very effective on poorer soils in the drier, eastern half of the country. These conditions are challenging for container-grown plants, particularly if the compost they are in is peaty. Once a peaty root ball dries out it is impossible to rehydrate it. Planting bare-root means you can add nutrients and organic matter (which holds air and water) and put your plant into the ground when dormant. If you keep it well watered in its first two or three growing seasons (April until late August), it will romp away. Bare-root trees are also cheaper to buy, and postage and packing is minimal.

SECRETS OF SUCCESS WITH BARE-ROOT TREES AND BUSHES

- Open the bag as soon as your tree arrives and check the contents.
- Plant as soon as possible – preferably in prepared ground.
- Do not plant in frosty conditions. Keep the bag in a cool, frost-free place, and if the roots look dry, add damp white paper kitchen towels to keep them moist. Alternatively, dig a V-shaped trench, then unpack the plants and lay them in halfway up the stems.
- Long roots can prove difficult when planting. Don't make a hole to fit the roots. Trim long roots back instead and then spread them out. The exception is the tap-rooted walnut (see page 266). Carefully back-fill, then lightly tread in the tree or bush.
- Trees are often grafted, but it's possible to see where the tree left the ground by looking at the trunk – just replant to that level. Order a stake with every tree.
- Water well.
- For further advice on planting, see January, page 7.

Organic Tip ✔

Don't overdo the organic matter. Add enough to make the soil friable, but make sure there's also plenty of garden soil. Add a sprinkling of bonemeal to the soil you're replacing: the phosphorous it contains promotes root development. Pre-preparing the ground allows it to settle so that the soil height stays in the correct position and doesn't sink.

2 Take Hardwood Cuttings of Gooseberries and Currants

(*early October*)

GOOSEBERRIES and currants are very easy to raise from hardwood cuttings at this time of year, although they do take their time to root. It will probably be this time next year before they are ready to move. They callous over in the first winter, then may produce roots in spring – lift them carefully during the following autumn to check. If rooted, either pot them up or move them to their final position; if there are no roots, give them longer – they will probably produce roots in the following spring. Very few cuttings fail.

Plant with the lowest bud 5cm (2in) above the ground, so that the currant or gooseberry bush will have a small trunk and bushy top. Rooting is most successful in a sheltered position and one of the best places is against a northern wall, or on the shady side of a hedge. Gardeners on heavy soil should line a trench with sand to aid rooting, though some prefer hormone rooting powder.

Gooseberries, redcurrants and whitecurrants can be trained as cordons and fans, and can produce a crop on a north-facing wall.

Victorian gardeners would grow a preciously early crop on a south-facing wall, followed by a later one on a north wall.

You can also raise figs and mulberries in this fashion.

Did you know? The jostaberry is a hybrid between the blackcurrant and gooseberry. It is thornless, with dark fruits that resemble gooseberries. These large bushes take 5 years to fruit and are not widely grown, although they make excellent jam. The jostaberry was developed in Germany and commercially released by the Max Planck Institute in Cologne in 1977. The 'josta' name comes from the German words for blackcurrant and gooseberry – *Johannisbeere* and *Stachelbeere*.

Organic Tip ✓

If you do not have too many cuttings, raise them in a large container of gritty compost, but keep them in the shade.

VARIETIES

For varieties of gooseberry, see January, page 13; for varieties of blackcurrant, see March, page 74.

SECRETS OF SUCCESS

- Select vigorous, healthy shoots from this year's growth.
- Dip the cut end in a hormone rooting powder to promote root formation.
- Plunge the cutting into the soil so that two-thirds are below the soil surface.
- Frost can lift the cuttings. Firm them in if needed.
- Do not let the cuttings dry out in the following spring and summer.

3 Protect Figs for Winter

(mid-October)

FIGS ARE sumptuous plants, but they come from much warmer climates than we enjoy in the UK. However, the National Collection holders (Reads Nursery in Norfolk, www.readsnursery.co.uk), with the help of their customers, have been researching which cultivars are the hardiest. My own very cold, windswept village (its name even begins with Cold!) harbours many thriving 'Brown Turkey' fig trees in the open ground. They crop well in warm summers, but always produce some fruit, whatever the weather.

Many gardeners grow figs in containers that can be moved into a frost-free position for winter, but if you do this make sure they are standing on pot feet to allow any winter wet to drain away. You can keep them in a greenhouse from August until late April; however, if a greenhouse is not available, a garage or garden shed will do between December and March – a shorter time, because a shed is much darker than a greenhouse. Keep the soil just moist and then pot on in March to a size about 5–7.5cm (2–3in) larger in diameter. Use John Innes

No. 3. Winter protection is advisable for figs planted in the ground – thick fleece is ideal.

Figs can be pruned in late March before growth starts. Cut out any dead wood or die-back to the healthy, white wood. Remove tips of young shoots and any thin, weak branches, just keeping the thick ones which are the fruit-bearers.

In warm climates a fig tree can produce three crops per year. However, in Britain figs planted in the garden will produce one crop each season and those planted in greenhouses will produce two crops in a sunny season. For harvesting figs, see page 235.

SECRETS OF SUCCESS WITH FIGS

- Provide a warm, sunny position and good drainage. Figs are fairly hardy, but the fruit needs warm sun to ripen.
- Restricting the roots – with wooden shuttering or paving slabs placed under the ground, for instance – minimizes the size of the tree and maximizes the fruit crop. Normally the bottom of the containerized box under the ground is left open to aid drainage but it is packed with a 22cm (9in) layer of rubble to stop large tap roots forming.
- Feed your figs with a general fertilizer in spring, then mulch them with well-rotted organic material.
- Once the fruits form, water on a potash-rich tomato feed until August.
- If you have restricted your fig, it may be thirsty. Water it well in the warm summer weather.
- If your fig drops its fruit in June, it is thirsty.

Did you know? Garden figs are parthenocarpic – the fruit develops without being fertilized. They live for centuries. Cardinal Reginald Pole, Archbishop of Canterbury, introduced the 'White Marseilles' variety to Lambeth Palace in 1525 and his trees are still flourishing.

Organic Tip ✔

If you do train your fig on to a south or south-westerly wall, add a stout support system to which you can tie it. Large-leaved fig branches can snap in gales.

VARIETIES

'Brown Turkey'
The most successful fig for cool climates. Reliable and popular, this mid-season variety produces a profusion of large, pear-shaped, dark-skinned fruits with dark red flesh.

'Brunswick'
Another very popular variety for outdoor culture in cool areas due to its hardiness. A mid-season fig bearing large fruits with yellowish-green skin and reddish flesh.

'Rouge de Bordeaux'
A gourmet fig for a very warm, sheltered site, or a conservatory or greenhouse. Deep-purple skin with red flesh.

'White Marseilles' (syn. 'White Genoa')
Attractive, pale-green–white skin with pale, almost translucent flesh. A good variety for growing outdoors.

4 Tidy Your Rhubarb

(late October)

RHUBARB leaves and stems are affected badly by frost and soon turn limp and soggy, it is a good idea to spruce up your rhubarb patch now, ready for winter. Remove or cut away any remaining leaves: they could be sheltering slugs and snails. Tidy up the patch now, and weed it, so that when spring comes you can enjoy the cooked stems of young rhubarb.

You can force dormant rhubarb crowns growing in the ground by covering them with purpose-made terracotta forcers or upturned dustbins full of straw. The dark, warm conditions inside force the rhubarb into growth a month early, causing it to produce soft, pale-pink stems that have a champagne flavour when cooked. However, once you have forced one crown it must be rested for 2–3 years and allowed to grow away naturally. Some gardeners just discard the crowns they have forced.

The simplest thing is to plant three crowns so that one is recovering from being forced the year before, one is cropping naturally and one is being forced. Then you will have a supply of forced stems every year. However, the yield from forced rhubarb is roughly half that of a plant grown outside.

Did you know? Rhubarb is technically a vegetable, but considered an honorary fruit. It needs a cold period before it can begin growing in the spring. Early varieties need a relatively short cold spell; later varieties a long cold period. If you want an early crop of forced rhubarb, you must choose an early variety.

VARIETIES

'Timperley Early' AGM (early)
So early it's probably better not to force it. The long, slender, pink-red stems have a tart flavour that makes it an excellent crumble-filler. Not a prolific cropper, but a must for all rhubarb-lovers.

'Hawke's Champagne' AGM (early–mid-season)
Delicately thin, long, scarlet stems with a sweet flavour from early spring. An old variety, but easy to grow and ideal for forcing. Attractive appearance.

'Queen Victoria' (mid–late-season)
Colourful, strong red stems, easy to grow and prolific. This heritage variety still holds its own today. Vigorous, forming huge clumps, so perhaps not for smaller gardens.

'Raspberry Red' (mid–late-season)
An old Dutch variety recently reintroduced. Heavy cropper for a sunny, open position. Sweet red stems.

SECRETS OF SUCCESS WITH RHUBARB

- Choose an open, sunny site and prepare the soil by working in plenty of farmyard manure or compost before planting.
- Plant in spring, where possible, placing new crowns 1m (3ft) apart with the buds just below the surface. Don't pull any stems until the second year of growth.
- Never cut rhubarb: the technique is to pull, then twist it very gently from the lower stem.
- Stop harvesting at the end of May to allow the plants to recover.
- If a stressed plant should run to seed, remove the flowering spike straight away. Water, feed and mulch lightly.
- Divide large clumps just as the dome-like buds are breaking dormancy. Lift the whole crown and, using a spade, split it into chunks containing four or five buds. Replant in enriched soil containing garden compost, making sure that emerging buds are just above the ground. Do not pick any stems in the first year or two and always remove flowering spikes.

Organic Tip ✔

Rhubarb prefers damp summer weather and a dry winter. Mulching with well-rotted compost or straw will help to achieve both, but do not cover the crowns deeply.

5 Grease-band Trees

(*late October*)

THE IDEA of applying grease bands is to prevent wingless insects (mainly wingless female moths) from climbing up the trees from the ground to lay their eggs. You can use them on apple, plum, pear and cherry trees. However, the sticky barriers don't work with Codling moths, which are the cause of maggoty apples. These fly into the trees in midsummer and you will need pheromone traps to upset them (see page 144).

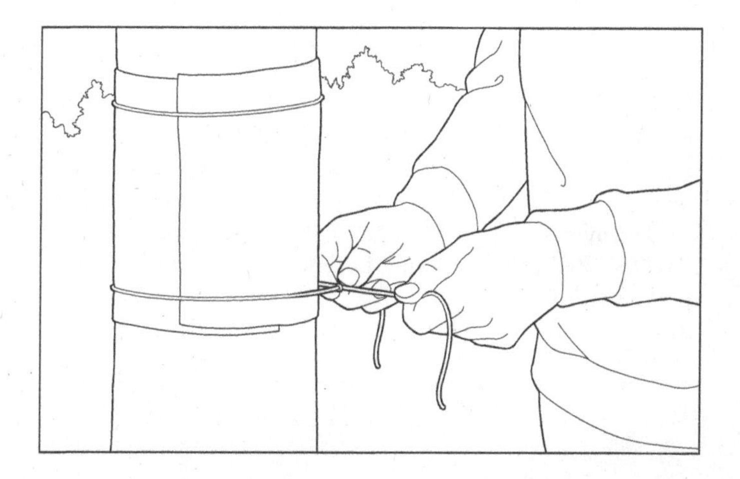

If you have a young tree with a smooth bark, use a ready-prepared sticky paper band – there are many on the market. If you have an old, gnarled apple tree, apply the grease directly to the bark – there are tins of fruit-tree grease and insect barriers on the market too. Place your band or grease about 45cm (18in) above the ground by late October, just before the adult moths emerge in November. Moth activity slows down in January, but wingless females are around until April. Re-apply the grease if needed.

Your grease should trap some of the wingless females before they reach the branches. The Winter moth, *Operophtera brumata*, is the most important. The adults are around between November and mid-January. When the wingless females climb up the tree trunk they make holes in the leaves and eat the blossom and fruitlets, affecting crop yields and quality. In early spring it's possible to see silken threads on the damaged leaves and by midsummer the leaves look very gappy.

The Mottled Umber moth (*Erannis defoliaria*) and March moth (*Alsophila aescularia*) eat leaves and fruit buds between late March and June.

Did you know? Around sixty species of butterfly are seen regularly in the UK, but there are around 2,500 species of moth. A hundred species fly in daylight, but most appear at dusk. They are just as affected by habitat loss as butterflies, with numbers dropping by a third since 1968. Some moths, like the Reddish Buff and Barberry Carpet, are highly threatened. Other species, like the Bordered Gothic, may now be extinct in the UK.

Organic Tip ✔

Don't spray! The tiny caterpillars of these moths form a large part of the diet of baby blue tits. They will clean up many of them. Blue-tit chicks alone feed on an estimated 35 billion caterpillars a year in Britain. The sharp drop in our garden bird populations is almost certainly related to the decline in moth numbers. They are an essential part of the food chain for birds, bats and mammals.

VEGETABLE

1 Cut Back and Weed Asparagus
(early October)

ASPARAGUS is a perennial crop that can stay in the same position for 20 years, so do look after it. Cut the ferny stems down once the foliage begins to yellow; this usually occurs after the first savage frost. Don't add the foliage to the compost heap: it may be harbouring asparagus beetle. Leave 2.5cm (1in) of growth showing, then weed the bed carefully by hand because asparagus is shallow-rooted. Add a thin layer of organic mulch, making sure you put it over warm soil. This will rot down and feed the crowns.

Mulching will also discourage seedlings from germinating next spring. Self-seeding asparagus plants can be a nuisance. If you spot any small seedlings, remove them all, as they will form inferior plants.

You can grow asparagus from seed, but this takes a year longer, so most gardeners choose to plant crowns. You will need at least thirty crowns to get a reasonable crop in the 6-week cutting season and these take up a sizeable area: crowns should be given at least 60cm (2ft) of space between each, with 75cm (2ft 6in) between rows. Plant them 7.5cm (3in) below the soil surface, then leave well alone until cutting begins in the third year.

Asparagus does best on lighter alluvial soil. If you have heavy soil that retains water, build a gentle mound (about 30cm/1ft high) to aid drainage, planting the crowns at the top of the mound.

SECRETS OF SUCCESS

- See page 193.

Did you know? Asparagus is almost certainly a plant from coastal regions of the Mediterranean and it thrives in light alluvial soils that warm up quickly in spring. Despite that, its range in the wild correlates to the old Roman Empire, indicating that the Romans planted asparagus as a staple, along with vines.

Organic Tip ✔

If planning to plant asparagus, weed the bed very thoroughly in the months before, removing all pernicious weeds. Being weed-free is essential for these shallow-rooted plants.

VARIETIES

For varieties, see June, page 194.

2 Make Leaf Mould

(mid-October)

THE ONE thing October has in abundance is leaves: one minute they cling to the branches, displaying their warm autumnal livery, and the next (usually after a gale-torn night) they are on the ground. Leaves are the source of a magical ingredient called leaf mould and this dark, friable mixture is adored by all plants whether spread as a mulch, added as a soil conditioner or used in potting up.

However, leaves can take up to 2 years to rot down thoroughly, so they cannot be added in huge quantities to a compost heap, although you can get away with small amounts. If space isn't an issue, build yourself a wire frame using chicken netting, then layer in the leaves and allow them to decay. Your dedicated leaf heap may take 18 months to deliver crumbly leaf mould, so you may have to bury next year's leaves underneath.

The leaves have to be damp when you collect them because dry leaves don't rot down. Certain species are slower at producing leaf litter: these include beech, oak and hornbeam. Others, like sycamore, horse chestnut and lime, are much quicker. Mowing over leaves first chops them up and then they rot down faster.

Many insects, amphibians and small mammals may be hibernating in the warm leafy layer under hedges and shrubs, so let these leaves remain *in situ*.

Did you know? Leaves rot down slowly with the help of fungi; the bacteria found in compost heaps do not do the job. These fungi prefer cooler conditions, so site your wire frame (or store your bags) in a cool position. Well-rotted leaf mould makes a good medium for sowing seeds.

Organic Tip ✓

You can make leaf mould using black plastic dustbin sacks. Collect the leaves and keep some air in the bag, then gently puncture the sides so that air can aid the decomposition process. Store the bags for at least 18 months, then tip the contents out and use the brown crumbly mixture in the same way. The mixture is wetter than leaf mould made in a bin, but still highly useful.

SECRETS OF SUCCESS

- Avoid evergreen leaves: they are too leathery to rot down.
- Pick up only leaves that are creating a nuisance – e.g. on paths, paving and grass, and on vegetable beds. Leave them to lie on flowerbeds and in quiet corners of the garden so that they rot down *in situ* and provide a habitat for small mammals and insects.
- Leaves on lawns can be chopped up and collected with a mower.
- Use a rubber rake and pick up the leaves by hand, using two small, light pieces of plywood to help.
- Water the leaves if they are dry.
- Transfer the leaves to your wire bin or black bags with your bare hands in case any hedgehogs, toads or frogs are hiding. Put anything back you find, choosing a quiet corner. Thick gardening gloves are too insensitive to allow you to feel hibernating animals. Take them off!

3 Remove Bean Canes

(late October)

IF YOU still have your bean canes in the ground, it's time to get them up and store them away from the wintry weather. Carefully cut any twine that might be binding the canes together and remove any vestige of climbing stems, etc. Using gloves, pull the canes out of the soil and lay them down on the ground. Always handle canes carefully. A bamboo splinter in the hand nearly always goes septic and great care has to be taken not to damage the eyes. Goggles would not go amiss.

Go through the canes carefully, separating out any with broken bottoms, etc. Shorten these with secateurs and store them for other uses. Keep all the long canes and tie them in bundles of ten, making a note of how many replacements you need to buy. An old chimney-pot makes a good storage container. Replace some long canes every year, either buying them now or in January. They sell out really quickly in spring and there are never any on offer in May – when you really need them.

Examine any pods. You may have some viable seeds and many gardeners do save their runner bean seeds from year to year. Bring them inside for a day or two to dry, then bag and label them before storing them in an airtight container in a cool place. The seeds are toxic, so do not leave them where children can get hold of them. If you are serious about saving seeds, elect to grow only one variety to lessen the chance of hybridization.

Organic Tip ✔

Bean seeds are viable for an average of 3 years once dried. The seeds are toxic if eaten raw, so always cook them thoroughly before eating them.

SECRETS OF SUCCESS WITH SEED-SAVING

- If saving seeds for a seed crop of any plant visited by bees, stick to one variety or separate them from other similar plants by a considerable gap.
- Runner bean varieties often have distinctive seeds, so discard any that differ from the majority.
- Green pods containing seeds can be dried in one piece before de-podding the seeds. However, if you remove seeds that are not fully ripe they will shrivel up and lose viability.

Did you know? Runner beans twist the 'other' way from most beans, climbing clockwise up the canes. Nearly all other beans turn the opposite way. They are also hypogeal: the cotyledons (the embryonic first leaves of the seedling) remain underground while the stem and true leaves emerge.

4 Begin Winter Digging

(late October)

WINTER digging sounds very onerous – but it isn't. All the gardener does is turn over the earth roughly with a fork. It can be done in any weather as long as you stand on a wooden plank to do it. This prevents the soil becoming compacted under your body weight. Two planks are better than one: then you can shuffle between them, moving one whilst standing on the other. Turn in any annual weeds: they will act as a green manure. However, remove any pernicious perennial weeds and any weeds with a seed head. These should be binned – don't compost them. Gardeners with heavier soil should turn it as early as possible because this is the most difficult soil for the weather to break up.

Using a large fork, up-end the soil once and it should form large clods. Resist the urge to break up the clods yourself. Leave them on the ground and their presence will increase the surface area of your soil. When frost occurs, the moisture in the soil will freeze and thaw, breaking up the lumps for you over 3 months or so. By the time spring arrives all you will need to do is rake through the soil and weed it. You will be left with a fine tilth, the perfect fine medium for sowing and planting. Jack Frost is an excellent ally.

Always leave the soil to settle after any digging, weeding or hoeing before you sow and plant, otherwise your seeds could be left on a ridge rather than on a plateau.

Did you know? Medieval gardeners left the weeds in place and ate the edible ones whilst they waited for their crops to mature. They also broadcast-sowed (sprinkled by hand) mixtures of three different crops, which meant that if one failed they still had food. These seed recipes were closely guarded secrets in the time of the Tudor market gardeners who fed London. Sowing single crops in rows did not become popular with gardeners until the mid-nineteenth century when the seed drill began to be used on farms, although Jethro Tull invented a simple seed drill as early as 1701.

Organic Tip ✓

Insect pests identify plants through receptors in their feet and they have to land on the same plant on four consecutive occasions before laying their eggs on it. Mixing up the leaves (as the Tudor gardeners did by sowing three crops together) must have made it harder for pests to target plants.

Insects also home in on the colour green and ignore brown, so a row of plants on the bare earth is an easy target. They don't identify the plant by smell. Scientists at HRI Warwick (Collier and Finch) proved that insects land on green paper as often as they do on green foliage, so the idea that aromatic foliage deters pests is almost certainly incorrect. However, they are confused by mixed plantings when several leaves mingle together.

SECRETS OF SUCCESS FOR SOWING

- When sowing seeds always use a line of string or twine to get straight rows. This saves space and makes it much easier to run the hoe up and down. Two lines are essential for accurate spacing. It's also useful to mark a 60cm (2ft) stick with measurements in centimetres, or inches and feet, which you can use to ensure your lines are parallel.

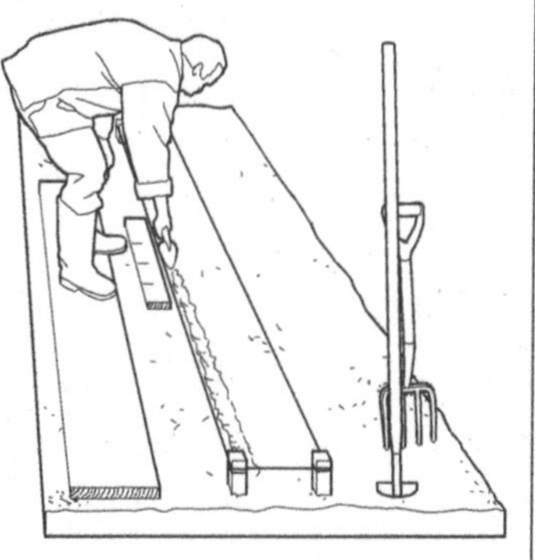

- Make a drill with a trowel or a rake handle. Water it if it's dry or if you are sowing papery seeds like parsnips.
- Sow thinly by tipping the seeds into the palm of your hand and then rubbing them through the fingers.
- Wide drills (about 10cm/4in in width) are excellent for crops like beetroot, parsnips and carrots because they save having to thin out the plants.
- Once the seeds are in the drill, the technique of 'shuffling in' is a useful one. Put one foot on either side of the drill and shuffle along the soil, making small movements. Don't pick your feet up. This will fill in the drill and firm the soil lightly.
- Cover newly sown seeds with chicken wire.
- Thinning plants (if needed) should be done in damp weather, as the seedlings pull out more easily. Cover up any resulting gaps with soil and then water the row again straight after thinning.

NOVEMBER

November is often the worst gardening month, because it's often damp, dreary and cheerless for days on end. Everything is in decline and decay. However, this lull in proceedings gives gardeners the chance to rest and recuperate after long hours outside, so use this window of opportunity to recharge your batteries and make plans for next year. Tidy the shed, surely the worst gardening job, give your garden tools a little bit of TLC (tender loving care), and make lists of sundries that need replacing. Once Christmas is over the garden centres will restock.

Browse through seed catalogues, while things are quiet, to see what's on offer, and order your seeds now so that you get the best choice. Vegetable specialists always offer the best value. Take some time to look back at the year, but be philosophical about your successes and failures. Gardening is always swings and roundabouts, and some years some things do well whilst others fail, despite your best efforts. That's life!

In the vegetable garden, it's a good opportunity to work on any empty plots when the weather is good. Dig over the ground with a fork, but don't break up the clods. Let the weather do this for you for because, as the ground freezes and thaws, the lumps break up. All you need to do in early spring is rake them over. You could also empty the compost heap, if it's well-rotted, and either spread it on your plots or double dig it in. Or you may prefer to use well-rotted manure, which is available in bags these days.

You should have winter crops to harvest, and these are akin to having a winter pantry because they'll give you months of food.

FRUIT

1 Erect Frames Around Peaches, etc.

(early November)

IT IS TIME to cover up peaches and nectarines to prevent peach leaf curl from getting into the buds. It is an airborne disease but, like all fungal diseases, warm and wet conditions encourage it. Putting up sheeting that covers the front of the whole tree from tip to toe will keep off the rain, making it harder for the disease to take hold if it's already present and also preventing new spores from landing on the tree when it's bare. It is the young leaves that are most vulnerable to the fungus: they distort, pucker up and redden, so it is very obvious that the tree is affected. The leaves fall off and then the tree becomes short of food and unfruitful.

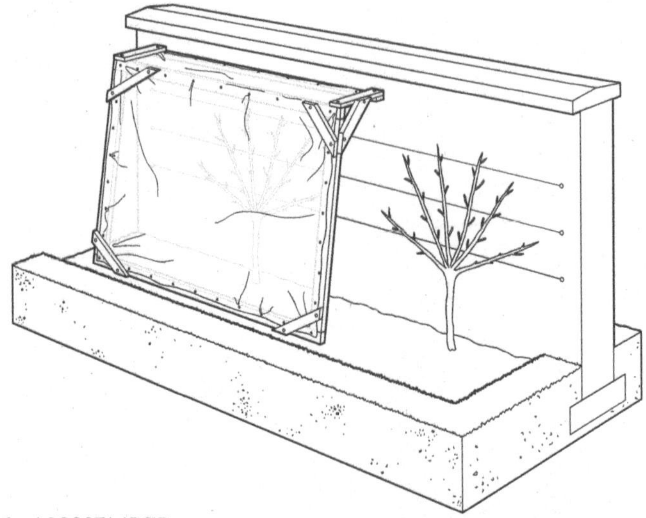

If this does happen, pick off any infected leaves and tidy any from the ground. Continue to do this. Always destroy or bin the leaves – never compost them. Mulch under the tree with bark to form a barrier between any spores on the ground and the tree.

Wall-trained trees are easier to cover up because you can install a pull-down sheet that covers the front whilst allowing air to enter and circulate from the sides. Free-standing trees can have tent-like structures put over them. Polythene is the best material because it isn't porous and it sheds water. It will also warm the area underneath. The sheeting should be kept in place until the fear of frost has passed – until early April is usually sufficient for a wall-trained tree. Lift it up to hand-pollinate your tree (see page 77). Porous fleece isn't a good idea, as a damp membrane exacerbates the problem.

Did you know? The closest relative to the peach is the almond – a ripe almond fruit looks very like a small, unripe peach. It has beautiful but early blossom that is often caught by frosts. Some named varieties of the sweet almond are available in Britain, but nut-set is often poor. If you do get nuts, the shells can be laughably hard – you may need a hammer.

Organic Tip ✔

Peach leaf curl appears most on outdoor-grown peaches; it is rare for a tree under glass to suffer. A seaweed-based feed will toughen up the foliage and make it healthier.

SECRETS OF SUCCESS

- For advice on growing peaches and nectarines, see April, page 116, and July, page 208.

VARIETIES

For varieties of peach resistant to peach leaf curl, see April, page 116.

2 Double Dig for Added Fertility

(early November)

YOUR SOIL is the most important component of your garden and many fruiting plants are long-lived, so the soil should be first rate. Compost heaps will be groaning and some may have fully rotted-down material that can be used in the garden. However, most garden compost heaps are not hot enough to kill off all the weed seeds and inevitably some will be left. Burying your compost underground lessens any weed problems and, whenever the weather is clement, it is possible to double dig areas and add your compost to them.

Double digging is easier to do on smaller areas and it is also much easier than it sounds. You need a groundsheet, a wheelbarrow, a spade and a fork. Lay out your groundsheet on the right if you are right-handed and on your left if you are left-handed. Dig out one spit of soil and pile it on the sheet to reveal a neat area. Then break up the bottom roughly with a fork, to a spade's depth if you can, to improve drainage. Pile your compost 10–15cm (4–6in) high and fork it in. Replace all your topsoil to form a barrow-like mound and allow it to settle. Plant it up in spring.

This fertile mound will double your yield of strawberries, or you can plant fruit bushes.

Did you know? Keeping off your soil is vital. The London market gardeners of the seventeenth century used long beds with paths on either side. Bed width was governed by a man's arm. They were never wider than two arms' lengths so that the gardeners could harvest from either side without having to tread on the ground. The vegetable garden at RHS Wisley has a section devoted to 1.2m (4ft) wide beds.

Organic Tip ✔

When you double dig you are boosting the soil's nutrients, especially nitrogen, but more importantly you are improving the soil structure. This encourages much better root development – and it is the root system that is the most important part of any plant.

3 Repair Wounds and Graft Trees

(mid-November)

APPLES and pears can be radically pruned in winter when the sap is not flowing. Peaches, nectarines, cherries and plums are normally only gently pruned in summer when the sap can seal the wounds – see page 208. However, occasionally large trees may need a damaged limb removing, and then a bitumen-based wound paint is normally applied. The tree will then form its own callus, so you need only treat it with wound paint once.

Care has to be taken when applying it: it could seal in dampness and cause more problems. For this reason opinion on its use is divided. However, most nurserymen and fruit specialists recommend wound paint, and you will certainly have to use it with cherries and plums to prevent disease from entering. Wound paint also prevents frost damage. When a branch snaps, leaving a large surface area, I recommend wound paint.

You can change the variety of apple you grow by over-grafting an established tree with the variety to which you want to change. You need to have scion wood – twigs from your chosen variety. Cut the scion into a 10cm (4in) length and make a slanting cut in it so that there's a 'tongue', or narrow sliver of wood, sticking out. Make a cut into the wood of your living tree, slip the two together and bind with raffia or grafting tape. Cover the join with grafting wax to exclude air. When the graft is successful, the new wood will take over.

Did you know? The Romans were adept at grafting and Roman soldiers were given plots of land to turn into orchards as an incentive to stay in Britain.

Organic Tip ✔

When working on more than one tree at a time, cut down the possibility of cross infection by washing your blades and cutting edges and then sterilizing them in bleach or boiling water. Rinse this off and dry well before cutting into the next tree.

SECRETS OF SUCCESS WITH CUTTING STEMS AND REMOVING LIMBS

- To cut stems, make the cut just above a healthy bud, or a pair of buds or side shoots. Where possible, cut to an outward-facing bud or branch to avoid congestion in the heart of the plant and to prevent rubbing branches.
- Do not cut too close to the bud, otherwise it may die. However, if you cut too far away you will encourage die-back above the bud and then rots and other infections may enter.
- When removing limbs, wear gloves and eye and head protection.
- Firstly, using a saw, make an undercut about 20–30cm (8–12in) from the trunk. Then saw downwards to prevent the bark tearing. This will leave a clean stub when the branch is severed.
- Secondly, remove the stub. Start by making a small undercut just outside the branch collar – the bump part where the branch joins the trunk. Make the overcut by angling the cut away from the trunk to produce a slope that sheds rain.
- Always avoid cutting flush to the trunk, as the collar is the tree's natural protective zone where healing takes place.
- If pruning cuts bleed sap, leave them to heal naturally – unless it's a cherry or a plum, in which case apply wound paint.

4 Tidy Up and Tie In Trained Fruit

(mid-November)

OFTEN THE autumn weather is mild and trees respond with a slight growth spurt. Have a good look at all trained fruit – espaliers, cordons, stepovers and fan-trained fruit – and prune out any growth that needs removing if it doesn't conform to the required shape.

Remember that heavily trained fruit is never heavily pruned. If it were, it would promote vigorous regrowth and that would spoil the shape. Most pruning takes place in August, when the growth spurt has slowed, so that any regrowth is subdued and contained (see page 232).

Make sure fan-trained trees are securely tied in and check the stakes and wires of cordons, espaliers, stepovers, etc. If necessary, loosen or replace ties.

Did you know? George Washington (1732–99), the first President of the United States, grew espaliers on his Mount Vernon estate in Virginia and one of his hobbies was pruning his apples. At the time his garden could have rivalled that of any French chateau. It is still in existence and still grows espaliers today.

Organic Tip

Don't apply a winter wash, even a so-called 'organic' mustard-based one. It will kill your predators. When tar-based winter washes became widely used in the 1920s they caused problems with mites (including the apple rust mite) because the predatory mites were also wiped out.

VEGETABLE

1 Sow 'Aquadulce Claudia' Broad Beans

(early November)

THE FIRST week of November is the traditional time to sow the hardiest varieties of broad beans (and early peas) straight into the ground. The seeds should start to germinate within 2 or 3 weeks, although in cold winters they may not appear above the ground until the New Year. Cover the rows to prevent mice and birds from eating the seeds and young plants. Chicken wire is usually adequate, although some gardeners resort to mousetraps in order to protect their peas and beans.

November sowings of both are a gamble. Seeds always do best in cold winters rather than warm, wet ones that stop and start. Sometimes the seeds fail to germinate in wet winters and rot in the ground. The young plants and emerging seeds can be eaten by mice, rats and pigeons and an autumn-sown crop will give you broad beans only 3 weeks earlier than early-spring sowings. Yet those 3 weeks are important because this crop can deliver in June when little else is available. November-sown crops grow slowly, but their root systems are strong and deep and generally the crop is heavy.

Did you know? Green mustard sprays, garlic washes, soft soap and dilute washing-up liquid sound a lot friendlier than insecticides and they may not linger in the environment like a branded chemical. However, they wreak just as much havoc in your ecosystem because they kill predators and pests indiscriminately. Remember your breeding birds need those insects, and if you feel a problem is getting away from you, attack the aphids or other pests with your fingers.

SECRETS OF SUCCESS

- You are in the lap of the gods at this time of year. For this reason, sow plenty of seeds and always cover them with wire. Remember the old adage: 'One for the mouse, One for the crow, One to rot and one to grow.'
- Zig-zag peas thickly across a 22cm (9in) wide trench. Add the wire and then the twiggy stakes immediately.
- Broad beans can be sown by dropping in one seed per hole to a depth of 5cm (2in) with 20cm (8in) spacings between them. Each pair of double rows should be 60cm (24in) apart. Sow a handful of seeds at each end of the row for gapping up any seeds that fail.
- If the November crop fails, send up a prayer and re-sow in late January. This is also the time to plug any gaps.

VARIETIES OF BROAD BEAN AND PEA

Broad Bean
'Aquadulce Claudia' AGM
The classic broad bean for November and early-spring sowing. This compact variety (90cm/3ft) produces large white beans.

Pea
'Douce de Provence'
Another round-seeded pea that can be sown in early November.

Pea
'Feltham First'
Round-seeded, dwarf variety (45cm/18in) for autumn sowing. Good cold tolerance. Pointed, straight pod 8cm (3.5in) long, with seven or eight peas per pod. This hardy pea overwinters consistently well and crops heavily.

Organic Tip ✔

Early aphid colonies are soon targeted by predators coming out of hibernation. So don't panic and spray them. If you do, the pests will bounce back quickly but the predators won't. You could end up with more pests, not fewer. You need those predators.

2 Make a Rotation Plan

(early November)

ROTATION plans involve growing crops in different places every year and are vital to all gardeners because they help improve yields. Crops exhaust the soil in different ways, so moving them on helps them to find the nutrients they need in fresher soil. Rotation also prevents a build-up of pests and diseases, especially soil-borne ones. If carrot root flies have been a problem, for instance, they should fade away if there are no carrots to feed on for 4 years.

Rotation plans can be incredibly complex, but I believe the simplest method is the best. A 4-year rotation works well. Divide your crops into the following categories – potatoes, legumes (i.e., peas and beans), brassicas and roots – and keep them in that order. The theory behind this is that potatoes exhaust the soil. The legumes replace the nitrogen used by the potatoes and this then feeds the brassicas that will follow on. Roots (and onions) come last and then every fourth year the plot is manured over winter. Confusingly, some root-forming crops are brassicas, including turnips, swedes and kohl rabi. These are subject to the same pests and diseases as other brassicas, so include them in the brassica section and not with the roots.

A 4-year rotation needs four equal-sized pieces of ground.

However, any practical gardener will know that the theory of sticking to a rotation plan is difficult, because you will squeeze crops that fall between the groups into gaps. These include sweetcorn and all cucurbits (squashes, courgettes, cucumbers, etc.). It is a yearly balancing act.

The important crops to move every year are potatoes, as they suffer from blight, eel-worm and scab. Root crops are prone to root fly and canker and they too should be strictly rotated.

SECRETS OF SUCCESS

- Make sure that every fourth year the ground is thoroughly enriched with well-rotted organic matter, either animal manure or material from your compost heap. This will improve soil structure and nutrients, and your soil will not dry out as readily.
- Use easy-to-apply sprinkle-on, organic fertilizers and liquid feeds to boost nutrients as appropriate and as recommended by the maker. Remember that these boost nutrients for a while but do not improve soil structure (see Fertilizers opposite).
- Over-feeding can result in soft, sappy growth that is prone to disease and insect attack.
- Remove all self-set potatoes as soon as they appear, as they could harbour disease and pests. They will also disrupt your pea and bean crops.
- Analyse problems and act accordingly. If potatoes continually get scab, look for a scab-resistant variety.
- Always select excellent varieties, preferably those with an award of Garden Merit (AGM).

Organic Tip ✔

Add fertilizers to plants only before and during the growing season, because these additions tend to be washed out of the soil during winter. Light sandy soils are usually hungrier. Clay soil is much more fertile, so needs less added fertilizer.

Did you know? Vegetable rotation was widely practised in Britain until the mid-nineteenth century when spread-on fertilizers became the fashion. Guano, the nitrogen-rich droppings of anchovy-fed sea birds collected from the coast of Peru, saw the rotation system's demise. Previously this odourless material had been used by the Incas. The first shipment arrived in Britain in 1842 and it soon became the country's most popular fertilizer, making fortunes for entrepreneurs like William Gibbs, who built Tyntesfield House near Nailsea, Bristol, for £70,000 – one year's profits.

THE EASIEST FERTILIZERS

Blood, Fish and Bone
An all-purpose, well-balanced, slow-release plant food for building fertility. Can be used when planting or sprinkled on later. Easy to handle and good for brassicas.

Bonemeal
A slow-acting source of phosphorus with nitrogen, best for developing good root systems.

Chicken Manure Pellets
Nitrogen-rich and easy to apply, but the smell can put gardeners off and attract the interest of foxes. Dogs will also consume the pellets. There are powdered forms too.

3 Plant a Native Hedge

(late November)

HEDGES are extremely useful to the vegetable gardener because they provide shelter from the wind and encourage and sustain birdlife. The end of November is an excellent time to plant bare-root whips to make a hedge. These arrive between now and early March looking rather like long sticks. They are inexpensive to buy (and post) and establish themselves quickly, as long as they don't have to struggle with weeds.

Native hedges contain species that have adapted to our conditions. They produce fruit and flowers that suit our wildlife, so they attract much more insect life than non-natives. These insects are vital to breeding birds – and birds are extremely good predators, so it's important to sustain them.

Did you know? Our native hawthorn (*Crataegus monogyna*) is the most appealing hedge of all to insects and can attract up to 149 species. The horse chestnut (*Aesculus hippocastanum*), which is an alien species, will attract only fifteen. So it makes good sense to plant a native hedge.

Organic Tip ✓

Source your hedging whips from a reputable supplier to ensure your hedge will flower and come into leaf at the correct time. Ask your supplier for information, as some hedging material grown in eastern Europe is not suitable in the UK. It may be the right species but the wrong ecotype: e.g., blackthorn from eastern European stock flowers far too early here to be of any benefit to our predators.

SECRETS OF SUCCESS

- Prepare the soil well before planting – preferably in the autumn. Dig a trench 45cm (18in) wide and 30cm (1ft) deep along the length of the proposed hedge. Improve the ground if you can by adding generous amounts of garden compost or well-rotted manure. If your soil is poorly drained, add sharp sand or coarse grit.
- If you want a thicker hedge, plant a double row about 60cm (2ft) apart.
- When the plants arrive, unwrap them and soak them for up to 2 hours in water. If the ground is frozen or waterlogged, before the plants arrive prepare a 'back-up' slit trench about 50cm (20in) deep in a sheltered place to rest your plants for safety before planting. Cover the trench with old carpet or polythene to keep the ground frost-free.
- Once planted, mulch well with at least 5–10cm (2–4in) of bark chips or other mulching material. This will suppress weed growth and retain moisture.
- Keep the young hedge well watered during its first growing season. In windy sites you may need to use windbreak netting.
- Leave the hedge bottom undisturbed, particularly in winter. The leaf litter will shelter amphibians, spiders, beetles, small mammals and insects. You may well have hibernating toads, hedgehogs and voles under your hedge.
- Cut your hedge in late winter. The best shape is an A, with the widest part of the hedge being at the base. This shape provides more shelter and the sides of the hedge get more sunlight.

VARIETIES OF NATIVE HEDGING

Hawthorn/Quickthorn (*Crataegus monogyna*)
Fast-growing, spiny deciduous plant tolerant of wet soils. Dark, glossy green leaves, clusters of prominent, scented white flowers in May followed by plentiful red haws in autumn. Very hardy and useful in coastal or exposed positions.

Blackthorn or Sloe (*Prunus spinosa*)
A dense, prickly deciduous plant. New shoots have a fine down but become smooth by winter, and are purple in sun or green in shade. Masses of snow-white flowers appear in March before the leaves, and these are followed by sloes, which turn from purple to black in autumn. Any soil, but will thrive on quite poor soil.

Dogwood (*Cornus sanguinea*)
A deciduous shrub with green stems flushed with red. Rich, damson-red autumn colouring. Any soil, very chalk-tolerant and thrives in a very damp position.

Spindle (*Euonymus europaeus*)
Green-stemmed deciduous shrub with inconspicuous flowers in May. Good autumn colour and the red and orange spindles attract birds, particularly robins. Spindle is a host plant for beet and bean aphids – so this is one you may want to avoid. Any ordinary soil, sun or partial shade.

Dog Rose (*Rosa canina*)
Arching branches bearing white to pale-pink, single flowers in June, followed by glossy red hips in autumn. Loved by birds in winter.

Hazel (*Corylus avellana*)
Deciduous shrub with large, mid-green leaves that appear with the bluebells. Long, yellow catkins in early spring are followed by edible nuts in autumn. Cut in winter for stakes and poles.

4 Set Up Water Butts
(late November)

IN THE days of the old walled kitchen garden there was always a dipping pond in the heart of the garden where you could go to fill up the watering can with rainwater. This precious liquid was at just the right temperature for the plants. Tapwater, which tends to be heavy on chlorine, is not nearly as good for mature plants as rainwater.

November, when there's less to do in the garden, is the perfect month to think about capturing some of that precious precipitation. Sadly, most of us have to catch our rainwater in a water butt positioned close to a shed or greenhouse. Most are not very exciting to look at, although you can order Grecian urn shapes and wooden barrels. Go for the largest you can comfortably site and make sure that it has a lid. This will prevent any accidents with children, small animals and birds. It will also stop mosquito larvae breeding.

Gardeners should not be using mains/drinking water for watering, except for seedlings. It is in short supply in many parts of the country and in years to come new-build houses will almost certainly be fitted with rainwater- and grey-water-harvesting systems as a compulsory feature.

Did you know? It was illegal to collect rainwater in the US state of Colorado until 2009. However, new research carried out in 2007 found that 97 per cent of the precipitation that fell in Douglas County in Denver never reached a stream. It was used by plants or evaporated on the ground. As a result of this finding, the ban on collecting rainwater was lifted.

SECRETS OF SUCCESS

- Make sure the tap on your butt is high enough to get a full-sized can underneath. Most water butts have a stand that supports them safely.
- If you live in a very dry area, invest in a square or rectangular tank. It will hold much more water.
- Order your tank from someone who can deliver it – they are not car-friendly items!

Organic Tip ✔

If 2.5cm (1in) of rain falls on 1,000 square feet of roof, it will yield 623 gallons of water. That would fill over 300 watering cans.

DECEMBER

It's better for the gardener if December is a chilly month, for your fruit trees and bushes need lots of chill days to promote good fruit buds. This is why Kent is a top fruit-producing area. Temperatures are low in winter, but spring and summer are usually warm and sunny.

Now that the leaves are off the trees it's a good opportunity to check for any damage, although the main pruning should wait until January or February. The golden rule is to prune on clement days when the weather's benign, otherwise the frost could penetrate newly pruned wood and kill it, causing dieback.

Winter vegetables should be in good supply, and leeks, parsnips, kale and Brussels sprouts can all be harvested in the run-up to the shortest day and in the days beyond.

Get into the habit of going out whenever the weather is reasonable. Continue to dig over any undug ground, provided it's dry enough. If the soil's too wet, concentrate on trimming the edges and weeding. Stand on a board to prevent your body weight from compressing the soil, if you have to. The really cold weather is more likely to bite in the New Year, but by then the days will be drawing out and that's a comforting thought for every gardener.

FRUIT

1 Winter-prune Grapes

(early December)

MOST VINES are sold in pots so they can be planted at any time of year, but they are best planted when dormant (November–March). Don't plant out a very small vine until spring, though, as hard frosts might kill it.

Vines are tolerant of a wide range of soil types, but they must have good drainage. Dig a hole deep and wide enough to take the roots when fully teased out. Double dig and add grit if drainage is a worry.

All vines need support, usually provided by a system of horizontal wires 30cm (12in) apart, beginning 38cm (15in) above the ground – low so that the vine benefits fully from the reflected heat of summer sun.

Vines are rampant plants intent on producing lots of leaves. They should rest between January and March and normally glasshouse grapes are kept as cold as possible at this time of year by opening the doors and ventilators. In April the greenhouse is allowed to get warmer.

If you are using the rod-and-spur system for a vine in a greenhouse or against outside walls (see page 171), after planting cut the vine back to two good buds above the graft point. In spring the best shoot – the 'heir' – is trained vertically to form the main stem. If disaster does not strike, the other shoot – the 'spare' – is cut back to a couple of leaves.

In the second winter the main shoot, or rod, is cut back by about two-thirds to two good buds, as are any laterals – the spurs – so that there is an heir and a spare at each point of the developing structure. You should end up with a main stem with very short

(2–5cm/1–2in) branches. In the third and subsequent winters, the rod is reduced by less – about half.

Also at this time of year the rod is untied and the tip is allowed to flop on to the ground. It is usually tied down horizontally in order to slow the sap and keep it away from the apical buds; this will also encourage lateral growth. When the buds begin to break, the rod is tied back into position.

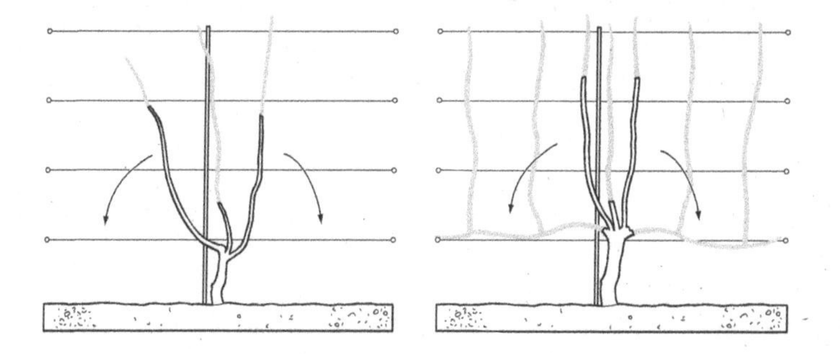

If your grapevine is very mature and rigid, or if it is outside, you may not be able to allow it to flop about. In this case just shorten the laterals as above and secure it well.

If you are growing vines in open ground using the double Guyot system (see page 171), the newly planted vine is again cut down to two good buds above the graft point. One shoot is allowed to develop vertically over the summer and any other shoots are cut back to one leaf. In the second winter the rod is cut back to three good buds at the height of the lowest wire. In the second summer the three shoots are again trained vertically and side shoots reduced to one leaf. In the third winter two of these shoots are cut back to about 75cm (2.5ft) long and they are trained along the lowest wire, one to the left and the other to the right. The third is cut right back to three strong buds for next year's replacement shoots. Your vine is now ready and set to produce fruit in its third summer.

Did you know? Tomato food is the best potash-rich food for fruit. It is widely available, but you can make your own high-potash food from 'Bocking 14' comfrey for nothing. Put the chopped leaves into a plastic container with a lid and allow them to rot down to produce a tea. Dilute it one part tea to twenty parts water. The drawback is that the decomposition process smells, although Garden Organic at Ryton use a wall-mounted drainpipe system which contains the stench. The pipe has a removable lid at the top and a tap at the bottom so that the leaves go in the top and the liquid is drained from the bottom.

Organic Tip ✓

A regular watering regime up until the longest day will prevent your grapes from splitting as they ripen. As the fruit forms and colours up, ease off the watering. If growing under glass, do not damp the floor down late in the day as the extra humidity encourages fungal disease.

SECRETS OF SUCCESS

- For further advice on growing grapes, see June, page 171, and August, page 237.

VARIETIES

For varieties of grape, see June, page 173.

2 Check Stored Fruit

(early December)

VERY FEW pears last beyond December, so it will be mostly apples that need checking now. If you have lots of stored fruit it's impossible to unwrap every apple unless you have lots of time. The best course is to check a wrapped sample in each tray or drawer. Gently feel the others, applying gentle pressure, for any signs of softness. If you've laid them out, well spaced, in wooden drawers, a quick glance will tell you if they are keeping well.

Check all your varieties and make sure that you are eating the ones with the shortest storage life first. If you feel that you cannot eat all the fruit that's ready, take some out of store and give it away.

Did you know? Over-ripe apples in store will give out lots of ethylene and this gas will cause other fruit – including later varieties of apple – to ripen precociously. Bruised fruit produces even more ethylene, so make sure you store only perfect apples. A single bruised one can wreak havoc.

Organic Tip ✔

Pick your fruit on the driest days that you can. It will keep for longer. Handle with kid gloves.

SECRETS OF SUCCESS FOR STORING

- Choose perfect, medium-sized fruit with stalks intact. Pick them slightly under-ripe for storage. Handle all fruit very carefully to prevent bruising.
- Use containers that will allow good air movement through the sides and over the top. Special wooden storage racks with drawers are available.
- Put fruits in a single layer so that they don't touch each other, stalk uppermost. Label them according to variety.
- Keep mid-season apples away from late-season ones so that they do not speed up ripening.

VARIETIES

For varieties of apple and pear suitable for storing, see September, page 260.

3 Tidy and Weed Fruit Cages

(mid-December)

FRUIT CAGES are a godsend: they ensure that you, rather than the birds, get your crop. However, we could get a heavy snowfall this month and the weight of overhead snow could easily bring down your cage or buckle the supports. At best, it will stretch and weaken the netting.

If you possibly can, untie the top, fold it carefully and store it away so that your cage is open to the elements. This will allow the birds access to the ground and they will make short order of eating any pests, including the raspberry beetle which overwinters as an adult.

This is also a good time to weed and tidy the ground under your fruit bushes and canes, as it is likely to have been trodden down every time you harvested. Weeds steal a lot of nutrients, so having a clear fruit patch will ensure that your fruit gets off to a flying start in spring.

Add organic matter now, or feed once the buds begin to break with a nitrogen-rich granular fertilizer. Replace the netting in April.

If you do not have a fruit cage, consider investing in one. The metal-poled ones are expensive – but so is fruit. If you are handy, it is much cheaper to make a wooden one and it will last between 5 and 10 years.

Did you know? Fruit cages seem to be recent arrivals, but in the nineteenth century the invention of barbed wire allowed the owners of cherry grounds to wire-fence their trees. Overhead netting was added to keep off birds.

Organic Tip ✔

Go to a specialist and try to get netting that's large enough for bumblebees to pass through to pollinate. Failing that, lift the sides in spring when much pollination takes place.

VEGETABLE

1 Clean Your Tools

(early December)

IDEALLY your garden tools should be cleaned after every use, but reality falls short of this expectation for most of us. My homage to daily cleaning is plunging the blades of my spade and fork into the bucket of damp sharp sand by my shed door. As long as the sand is damp it will keep the blade clean and sharp. Hand trowels and forks get the bucket-plunging treatment too.

But now it is worth having a blitz on everything in the shed, from rakes to brooms to shears. Arm yourself with newspaper and old cloths, buy some linseed oil and get busy. Wipe all wooden handles with a damp cloth and dry them before oiling the handles liberally. If the wood feels too rough, sand it off beforehand. An oily cloth can also be wiped over the metal to prevent rust. Store your tools off the ground, placing rakes and brooms head up. Check the wheelbarrow too.

This yearly clean makes it a joy to handle tools the following year. Most of us are devoted to our favourite fork or spade and keeping it in fine fettle is important. However, if replacements are needed this is an excellent time of the year to break new ones in, because you have more time to get used to them.

Tools make excellent Christmas presents. I particularly admire stainless-steel spades, but I always handle a garden tool to make sure the whole thing feels balanced before buying or dropping a strong hint for Christmas. Good secateurs are essential too, though very expensive. A good pair should last a lifetime.

If you have a mower that needs servicing, get it sorted out now before spring arrives.

Did you know? The wheelbarrow is believed to have been invented by the Chinese general Chuko Liang (AD 181–234), who used wooden wheelbarrows to transport supplies and collect injured soldiers. His wheelbarrow had two wheels and required two men to propel and steer it. The earliest evidence of a European (or single-wheeled) wheelbarrow is found in a stained-glass window in Chartres Cathedral in France and dates from 1220. The wheel is near the front to provide enough leverage to move heavy loads.

Organic Tip ✔

Be sustainable where possible and use oak labels, metal cans and jute string. They might cost more, but the joy of having materials that blend into the garden is worth the extra cost.

2 Tame the Snail

(early December)

ONE OF the most valuable things you can do now is go on a snail hunt for hibernating clusters. You'll find them against sheltered walls and behind plants. Having sealed up their shells, they hug together in clusters that can be as large as a football. If you can locate them now, before they wake up in spring, you can save yourself a lot of trouble. Check under water butts, close to sheds, in the greenhouse and in any sheltered nooks and crannies.

Snails are not slimy to handle at the moment and if you crush them hungry thrushes and blackbirds can reap the benefits. Once spring and summer arrive, snails will be on the move throughout the day. Lay small plastic pots on their sides and check them daily for sheltering snails. They also like to slide up and down linear leaves, so you often find them on irises, pokers and hemerocallis. Learn to frisk susceptible plants regularly – wear rubber gloves if you're squeamish.

If you squash some snails on the path as dark falls in spring and summer, I can guarantee that slugs will appear to devour their remains. Slugs are creatures of the night, coming out as darkness falls to seek prey. That's the best time to hunt them down.

Did you know? The common garden snail (*Helix aspersa*) is the fastest species of snail. It can move about 55m (60yd) per hour. They are also very strong – they can lift ten times their own body weight. They hate bright sunlight. Snails sniff out their food despite the fact that their two upper tentacles have eye-like light sensors. The shorter two are for feeling.

Organic Tip ✔

The song thrush relies on the snail as a staple food and is able to smash the snail's shell against a stone, so having some flat pieces of rock in your garden is a great help to them.

3 Sort and Order Vegetable Seeds

(mid–late December)

TIME TO sort out your old packets of seeds and decide what to buy for next year. Seeds that have a really short viability are parsnips (2 years), carrots (3 years) and sweetcorn (2 years). Check the dates of these carefully. Many seeds (including runner beans, beetroot, cabbages and lettuce) can store for 5 years in perfect conditions. However, open packets can deteriorate and I find lettuce seeds have a tendency to do so.

When ordering, it's often worth investing in F1 vegetable seeds, although they are expensive and you always get fewer seeds. However, they have hybrid vigour and produce a crop enthusiastically. Their seeds also germinate more readily, so F1 varieties of sweetcorn and parsnip (both troublesome in their own way) are well worth the investment, as is F1 spinach.

In recent years vegetable-breeders have risen to the organic challenge by breeding disease-resistant varieties, so seek these out. Look out for eastern European varieties too: these are often very hardy, prolific and disease-resistant because chemicals were not widely used in eastern Europe, as they were too costly.

There has also been a breakthrough in bean-breeding and there are new French × runner bean crosses. The first was 'Moonlight', a bean that looks like a cream-flowered runner. The French bean blood allows it to crop for longer (even in hot weather) and the individual beans are plumper – see page 159.

Balance the expense of F1 varieties by ordering well-established, top-performing AGM varieties like pea 'Hurst Greenshaft' and broad bean 'Jubilee Hysor'. Thompson & Morgan have an AGM section. If you have a variety that works for you, stick to it.

Did you know? In dry conditions seeds store for a surprisingly long time. A Judaean date palm seed recovered from excavations at Herod the Great's palace at Masada in Israel was carbon-14 dated to 2,000 years old. However, it still germinated in 2005. In 2002 Chinese archaeologists discovered an ancient fruit cellar in Shaanxi Province containing well-preserved apricot and melon seeds estimated to be 3,000 years old. After 8 years of research, they concluded it was a cellar used to preserve fruits for aristocrats.

4 Force Witloof Chicory

(mid–late December)

WITLOOF chicory is a very useful crop because you can lift the roots from late November onwards and force a few roots at a time. This provides a vegetable during winter when the ground may be frozen or even covered in snow. The crunchy, pale leaves are good either cooked as a vegetable or eaten raw in a salad – a reminder of summer past.

Lift four or five roots and use scissors or shears to cut off the tops to within 2.5cm (1in) of the top of the rootstock. Place them in a large pot full of compost so that the tops are proud of the soil. Cover them with a bucket so that they are in the dark, then leave them in a cool place (average temperature 16–18°C [61–64°F]) for 3–4 weeks. In that time a pointed sheath of crispy, pale leaves will appear: this is known as a chicon. You can slow the process down by moving the pots outside if you wish.

You can force other chicories and endives, or you can use them for leaf. However, the seeds are sown at different times. Cultivars of chicory (*Cichorium intybus*) for forcing are sown in May and June. Leafy varieties are sown a month later, in June and July. Endive

(*Cichorium endivia*) can be sown from April to August. Sow thinly to a depth of 0.5cm (0.25in) in rows 30cm (12in) apart.

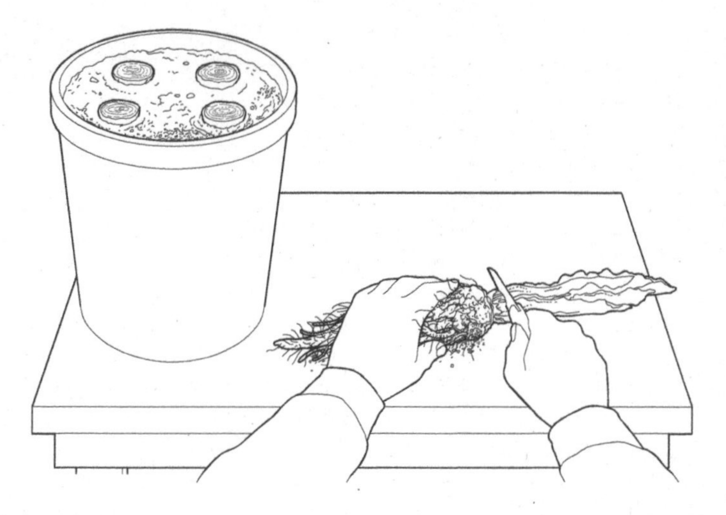

Did you know? After the discovery of blanched chicory in Belgium in 1830 (see page 163), a horticulturist from Brussels Botanical Garden refined the forcing process and chicons were first sold in Brussels market in 1846. However, forced chicory did not become widely eaten until the 1920s and 1930s.

Organic Tip ✔

Don't sow the seeds for forcing too early in the year, otherwise your plants could bolt. Thin the seedlings carefully so that they form large roots and then you will produce large chicons.

SECRETS OF SUCCESS

- Either sow the seeds directly into open ground in late May or early June and then thin them out so that each rosette is at least 15cm (6in) apart, or raise in small pots and bed out in late June.
- Keep the plants weed-free and water well in dry conditions.
- Lift the roots, then cut back the foliage to within 2.5cm (1in) of the crown in early November.
- Store them horizontally in peat or sand in a box in a cool shed or garage.
- Force a few of the roots at a time. Place two or three in a 22cm (9in) flowerpot so that the roots are just above the soil. Place an upturned pot or black bucket over the top. Leave in a dark place where the temperature is around 16–18°C (61–64°F).
- Keep the pot of compost or soil moist and warm. The blanched endives or chicons will be ready to harvest in about 3–4 weeks.
- When harvesting a chicon, cut into the rootstock or the chicon will fall apart.

VARIETIES

For other varieties of chicory, see May, page 162.

INDEX